AF587358

COMPUTER NETWORKS SERIES

FILE-SHARING APPLICATIONS ENGINEERING

COMPUTER NETWORKS SERIES

File-Sharing Applications Engineering
Luca Caviglione
2009. ISBN 978-1-60741-594-7

COMPUTER NETWORKS SERIES

FILE-SHARING APPLICATIONS ENGINEERING

LUCA CAVIGLIONE

Nova Science Publishers, Inc.
New York

For permission to use material from this book please contact us:
Telephone 631-231-7269; Fax 631-231-8175
Web Site: http://www.novapublishers.com

NOTICE TO THE READER

The Publisher has taken reasonable care in the preparation of this book, but makes no expressed or implied warranty of any kind and assumes no responsibility for any errors or omissions. No liability is assumed for incidental or consequential damages in connection with or arising out of information contained in this book. The Publisher shall not be liable for any special, consequential, or exemplary damages resulting, in whole or in part, from the readers' use of, or reliance upon, this material.

Independent verification should be sought for any data, advice or recommendations contained in this book. In addition, no responsibility is assumed by the publisher for any injury and/or damage to persons or property arising from any methods, products, instructions, ideas or otherwise contained in this publication.

This publication is designed to provide accurate and authoritative information with regard to the subject matter cover herein. It is sold with the clear understanding that the Publisher is not engaged in rendering legal or any other professional services. If legal, medical or any other expert assistance is required, the services of a competent person should be sought. FROM A DECLARATION OF PARTICIPANTS JOINTLY ADOPTED BY A COMMITTEE OF THE AMERICAN BAR ASSOCIATION AND A COMMITTEE OF PUBLISHERS.

Library of Congress Cataloging-in-Publication Data
Available upon request.

ISBN 978-1-60741-594-7

Published by Nova Science Publishers, Inc. ✤ New York

Contents

Everything of worth
On Earth
Is there to share
***Morrissey**, Glamorous Glue*

Acknowledgments

I am deeply grateful to people who helped in reviewing and improving this book. Without any particular order, I would like to thank Franco Davoli, Cristiano Cervellera and Michele Sciuto.

Besides, I would like to hug my family (Gianni, Grazia and Berto) for supporting my efforts. In addition, friends and colleagues are always a precious source of inspiration and support. Many thanks to: Giuseppe Spanò, Davide La Rosa, Armando Barbano, Thomas Moriconi, William Yeager, Ludovica Tedeschi, Tomaso de Cola, Matteo Balicco, Sandro Scotto, Carmelo Marullo, and Silvia Maldina.

Maria Assunta Marseglia now is my happy thought.

Also, I am grateful to people at ISSIA (Genoa Branch) for the great support (Filippo Aldo Grassia, Roberto "Bob" Marcialis, Marta Cuneo and Elisabetta Punta).

Lastly, I would like to thank Andrea de Marco and Dario Tortora: they were the firsts trusting in my writing abilities.

About the Author

Luca Caviglione got the Laurea Degree in Telecommunications Engineering from the University of Genoa (http://www.unige.it) in 2002. In 2006, he obtained the PhD in Electronic and Computer Engineering from the same University. From 2006 he is a Contract Professor at the University of Genoa, within the course of Telecommunication Networks, in the fields of Multi-Service Satellite Networks, Switching Systems and peer-to-peer (p2p) Networking. He also taught in several Master courses, lecturing about p2p and file-sharing applications.

He participated, and still participates, in many Research Projects funded by the European Union (EU), by the European Space Agency (ESA), by the Italian Ministry of Research (MIUR) and by Siemens COM AG.

He is author and coauthor of more than forty academic publications (conferences, journals and book chapters) about TCP/IP networking, p2p systems, QoS architectures and wireless networks. He participates in TPCs of several conferences, regularly serves as a reviewer for some major journals, and gives talks about the usage of IPv6 to empower p2p applications.

In 2006 he was with the Genoa Research Unit of the Italian National Consortium for Telecommunications (CNIT, http://www.cnit.it). Since 2007, he works at the Genoa Branch of the Istituto di Studi sui Sistemi Intelligenti per l'Automazione (ISSIA, http://www.issia.cnr.it) of the Italian National Research Council (CNR, http://www.cnr.it). He is a Work Group Leader of the Italian IPv6 Task Force (http://www.it.ipv6tf.org) and he has filed, as a coauthor, several patents in the field of p2p. Moreover, Luca is involved in the Satellite Communication Network of Excellence (SatNex, http://www.satnex.org), where he studies the impact of p2p communications over satellite systems and viceversa.

Luca is also a Professional Engineer and a technical writer for the most popular Italian Macintosh magazine.

He can be reached at: luca.caviglione@ge.issia.cnr.it.

Preface

I met Luca many years ago, we were both PhD students eager to work in technology and to spend our time playing with p2p networks. That time has gone, beaten by funny talks, laughs and reflection (and, of course, not just by source code and network cards). Nowadays we both left University, but fortunately he is still in research, driven by passion and curiosity for his work.

I read this book with interest and fun: to the best of my knowledge it is the first complete scientific publication about p2p file-sharing networks, and, thanks to the author, it does not comply with the traditional scientific literature. It follows a light wire able to bring the reader deeply into the core of p2p technologies, missing boring scholastic surveys and traditional introductions. I would say the text respects the personality of the author, which could be really defined, somehow, a storm of unbounded emotions and illuminated rationalism.

Now, it might be you cannot understand why I am writing such things, but the reason will be clear soon: this book is what you could desire before starting to read a publication about technology, a mix of passion and logical thoughts. The reader will find out how it can be pleasant to get involved in technology if the writer is able to transmit his curiosity and deep knowledge for the proposed topic.

Michele Sciuto

Organization of the Book

This book offers a self-contained discussion about file-sharing systems from an engineer's point of view. Its main scope is to rationalize the engineering process at the basis of peer-to-peer (p2p) file-sharing systems, also in the perspective of actively developing this type of network applications. The work analyzes the architectural blueprints, the design choices, the internals, the core algorithms, their interaction with the underlying network infrastructure and some of the major findings of the scientific community. With such foundations, it will be straightforward to understand the main behaviors of these systems, their strengths and weaknesses and to correctly evaluate the resulting traffic patterns. In addition, to complete the picture, source code analysis, techniques to profile the deployed services and simple examples/exercises are provided.

As a consequence of the widespread diffusion of file-sharing applications, of their impact over the network in terms of traffic load, and of the kind of content conveyed, different parties are interested in the topic. This book offers a comprehensive analysis of file-sharing applications dealing with different aspects of the same technology, thus accomplishing various needs of knowledge with a different degree of complexity. There are several reasons to read about file-sharing applications even if some readers' goals clash with others'.

Summing up, the **intended audience** is:

- academics and researchers: many file-sharing services are populated by several millions of users simultaneously, representing a challenging research environment, to understand, characterize, model and experiment with. This book can be used as a focused introduction to the topic or for quick reference purposes, i.e., to get a deeper insight on major client interfaces and their internals;

- students: file-sharing is often used to portrait to students real examples of successful p2p applications. Moreover, due to its challenging use of network services, implementing a basic file-sharing application has pedagogical purposes to understand how to write good network-oriented code. In this case, the book can be used as a comprehensive text-book on the topic;

- network administrators: the resulting heavy traffic loads, jointly with marginal responsibilities of the Internet Service Provider (ISP), reflect in a daily challenge for network administrators. Via a thorough explanation of the design choices and the interaction paradigms, network administrators can use the book to understand how to develop and deploy proper countermeasures;

- businessmen: file-sharing services highly rely on end users cooperating and sharing their own resources. As a consequence, such applications present new business-models also giving the opportunity of starting new services at very low costs. The book portraits the main technologies adopted for developing effective file-sharing systems, thus providing a guideline to better understand the power of user-based co-operative distributed architectures.

A discussion on file-sharing technologies may rise some issues about **legal aspects**. This is due to three significant matters, namely:

- the majority of the exchanged material is copyrighted: the popularity of file-sharing applications among users and early adopters is due to the availability of a rich collection of music, movies and software. Even if the actual trend is to rely on the efficiency of modern file-sharing systems to distribute software while flattening or zeroing costs[1], the presence of "illegal" material still plays a key role for their appeal;

- the highly distributed flavor of modern file-sharing architectures is used to hide crimes: tracking a file or a resource consumer in a centralized system is simpler than in a highly distributed one. For this reason, some aberrant crimes, such as distributing child pornography, take place in some p2p file-sharing services;

- many Nations are developing Laws to regulate users' responsibilities when exchanging copyrighted material through the Internet: even if this action may clash both with the freedom of the Internet and of individuals, it puts pressure over end-users, application developers, ISPs and network administrators. In this vein, traffic monitoring and event logging are becoming crucial, not only for avoiding to endanger Quality of Service (QoS) disciplines or network performances due to traffic swamping.

However, the book will not deal with "ethical" aspects, but it will analyze some major engineering choices to cope with "environment" aspects. In addition, as it will be discussed later, the "legal" aspects have somehow always driven the development and the adoption of new techniques, such as relying over more distributed architectural flavors. As a consequence, some "countermeasures" will be presented from a neutral engineer's point of view, without further comments. Readers interested in legal aspects can find additional details in [1]. In a word, this book is **neutral** with respect to the file-sharing phenomena.

Book Focus and Possible Overlaps

In order to maintain the discussion focused and short, we decided to deeply analyze only two file-sharing systems: eMule and BitTorrent. As today, such applications are both the most sophisticated and adopted ones. Besides, other major systems, such as Gnutella will be investigated, even if with a lower level of detail. Concerning insights useful to engineer and

[1] For instance, many software developers are delivering Linux distributions through the BitTorrent application, which will be analyzed in Chapter 7. In addition, user-based initiatives, such as developments of customized on-line games, rely also on seeding through the p2p networks the produced software, which is typically in the order of hundreds of Mbytes.

analyze file-sharing applications, specific tweaks and solutions adopted in other systems will be discussed when needed.

Modern client interfaces, even if based on very different design patterns, are progressively converging over some technologies and functionalities. For instance, the usage of distributed networks to locate contents and to recover from failures of centralized components, as well as employing fine grained algorithms to manage files. Thus, in order to avoid overlaps, we will try to describe in a "mutually exclusive" manner the features of the two major systems we will present later in the book (i.e., eMule in Chapter 6 and BitTorrent in Chapter 7).

Outline of the Book

The rest of the book is organized as follows:

- **Chapter 1** introduces at a high-level of detail the notion of file-sharing application and a quick historical revision of the phenomena, which culminated in the adoption of the p2p architectural blueprint;
- **Chapter 2** offers an introduction to the p2p communication paradigm focusing on aspects strictly related to their adoption to engineer file-sharing applications;
- **Chapter 3** analyzes both the basic components and the core functionalities needed to implement a successful and effective file-sharing application;
- **Chapter 4** presents a portrait of the major pitfalls introduced by the underlying network architecture. Specifically, it addresses problems arising due to the lack of transparency in the modern Internet and it deals with the most popular traversal techniques;
- **Chapter 5** proposes an analysis of the most popular heuristics employed to force file-sharing users to play an active role within the overall service;
- **Chapter 6**, with the foundation of the previous chapters, showcases the eMule file-sharing system.
- **Chapter 7** discusses the BitTorrent file-sharing system, and it introduces the concept of service capacity of p2p file-sharing and content replication frameworks;
- **Chapter 8** deals with a survey of the most recent advancements achieved by the underground community in developing file-sharing applications and of the main modifications and enhancements proposed within the scientific literature. Besides, it also deals with the optimization of p2p file-sharing systems. Such topic has been also introduced to provide examples on how to model some critical parts of real applications;
- **Chapter 9** offers a survey of the literature about the traffic analyses of file-sharing applications. In addition, it introduces methodologies and tools useful to gather and understand patterns produced by such services;

- **Chapter 10** investigates the reference implementation of the applications discussed in Chapters 6 and 7 (i.e., eMule and BitTorrent), analyzing the source code, the protocol implementation and the possible modifications to enhance performances, gather data and develop instrumented client-interfaces to conduct research;
- **Chapter 11** proposes the conclusions and summarizes the content of the book. In addition, it presents some possible exercises to better understand concepts explained in this book.

Before starting the discussion, we would like to cite the words of Dr. Vinton G. Cerf, who is one of the Fathers of the Internet, along with Dr. Robert E. Kahn:

"*People who think peer-to-peer is a new idea should understand that when we were designing the Internet, that was a critical and core part of the design. The TCP/IP protocols are precisely designed so that all elements in the Net are essentially of* ***equal*** *standing*" [2].

Part I

FUNDAMENTALS

Chapter 1

Introduction to File-Sharing

This chapter offers a concise introduction to file-sharing applications, which allow to share and exchange files among remote users through the Internet. Such applications have a quite short, but fast-moving history. As today, file-sharing services are used by millions of users on a daily basis, and they are responsible of generating a very relevant fraction of the overall traffic of the Internet [3]. They represent an important set of network applications and, as a consequence, they are under the attention of several actors.

Besides, the genesis of file-sharing applications is quite interesting, since it represents a revolution started "from the bottom". Even if modern systems are developed also by taking into account ideas borrowed from the literature, the first impromptus came up from non-professional developers or engineers.

Then, in a nutshell: *file-sharing is substantially a product of the underground populating the Internet.*

The chapter deals with a brief introduction to what a file-sharing application is, a quick historical perspective (emphasizing the importance of Napster as a catalyst of the phenomenon) and quick overview of the most popular services available through the years.

1.1. File-Sharing Applications in a Nutshell

File-sharing applications (rarely called also *file-swapping* applications, as to emphasize the file movements among different entities) allow a user to host **files** on his/her computer and then to exchange them through a telecommunication network, mainly the Internet. A given host running a proper software, often called *client interface*, is responsible for receiving and transmitting files, acting both as a service consumer and producer.

Despite the communication paradigm, this user behavior is tightly coupled with the peer-to-peer (p2p) model, where hosts act both as a client (receiving a content) and a server (being responsible also of the content delivery). The word **sharing** emphasizes the "constant" availability of files at the user side.[1]

[1] Typically, files to be served are collected into a specific folder called the *shared folder*. Obviously, multiple folders can be shared simultaneously. In order to locate files to be exchanged, search functions are also present. Due to the duplex role of users, some systems also utilize mechanisms to promote cooperation and to avoid egoistic behaviors. Such mechanism will be investigated in Chapter 5.

1.2. Is There a Philosophy Behind File-Sharing?

The usage of file-sharing applications has increased dramatically during the last years. A cursory analysis shows that the traffic load generated by such applications is decreasing due to the menace of legal threats. But a more thorough look would reveal that this is not actually true. In fact, it appears that the file-sharing traffic patterns are mutating and hiding in the Internet's "mare magnum" rather than decreasing [4]; a discussion regarding traffic dissimulation techniques will be presented later in the book.

Summarizing, spending time with such programs is a habit for many users, and it is not uncommon for people to have a computer connected 24/24 h to some exchange network, or using (if present) its integrated Instant Messaging (IM) subsystem to orchestrate the exchanging effort. This to say that many file-sharing application users perceive this as a "philosophy" of the Inernet usage.

Lastly, consider the next phrase, taken from the must-read RFC 3751 Omniscience Protocol Requirements [5]: "The Dark Side of a Bright Idea: Could Personal and National Security Risks Compromise the Potential of Peer-to-Peer File-Sharing Networks?". Even if RFC 3751 has been written as the traditional April-fool standard document, it somehow underlines the attention of different communities to file-sharing networks.

1.3. Napster

The first remarkable attempt to bring this technology to the masses was Napster, developed in 2001 by Shawn Fanning and aiming at offering a comprehensive platform to exchange multimedia audio files. Fanning's idea was simple yet revolutionary: instead of using a central repository to deliver files, users are allowed to directly exchange data. Napster was not completely based on the p2p communication paradigm and it relied on a centralized component to maintain the service running, but it was unique at that time. Napster grew fast, and focused the attention of the recording industry[2]. However, due to its critical centralized component responsible of orchestrating users searching for files, stopping the overall system is fairly simple. In fact, after a long legal epic, Napster was shut down and, later, its brand used for commercial purposes. Napster inspired many developers, while its struggle emphasized its design flaws, but started the massive development and diffusion of file-sharing applications.

Napster was a mainstream owing to the superposition of three different main aspects:

- music had, and still has, a great attraction for the masses. In addition, Napster represented a quite different application for the average Internet user, often employing only e-mail, web browsing and Internet Relay Chat (IRC): in a nutshell, it was a real killer application;
- the advent of the mp3 (MPEG-1/2 Audio Layer 3) compression algorithm enabled to exchange high-quality audio contents over the widespread Plain Old Telephony Service (POTS) dial-up accesses. Years later, the same gap was filled in the video encoding by the DivX;-) compression algorithm;

[2]Details on the Napster's struggle against the Recording Industry Association of America (RIAA), which was instigated by Metallica, has been discussed so many times that it has been intentionally omitted here.

- the system was designed to partially exploit a disruptive technology: the p2p communication paradigm.

Despite the first reason, that is "emotional" rather than technical, the other two are important engineering achievements needed to by-pass some technological limits imposed by the network environment available at that time[3].

The wide adoption of the mp3 technology allowed to partially overcome bandwidth shortage by dramatically reducing the dimension of audio files, bringing their transmissions over POTS or basic and narrowband Integrated Service Digital Network (ISDN) Internet accesses feasible. The third factor appeared as critical for the effectiveness of data exchanging among users: the p2p communication paradigm. In fact, replacing a portion of the classical client-server architecture with a p2p one, also allows to overcome some bandwidth limitations at the user side, while relieving the need of employing a large set of resources at the server side.

Then, this lesson was important for p2p "hackers": after the Napster's attack, its successors were based on a more distributed flavor.

1.4. Pre-Napster File-Sharing

Prior to the advent of applications solely dedicated to file-sharing purposes (e.g., the aforementioned Napster), the file exchange activities were already present. For instance, users exchanged material through newsgroups, by posting messages with binaries attachments. For the interested readers, a popular hierarchy allowing such operation is still today the *alt.bin.** one. This process has some difficulties and it is extremely fragile, since: i) the file to be delivered must be "chopped" into several small parts; ii) each part must be posted into a separate news message as an attachment; iii) to reconstruct the original file, all the messages must be retrieved and their attachments subsequently merged. Obviously, a missing news in the server (e.g., it has been removed by the administrator or its lifetime has expired) will compromise the reconstruction of the original content. As to architectural requirements, a Network News Transfer Protocol (NNTP) server must be in place to enable users to post and retrieve messages. Besides, the application utilized to manage the news (called the *newsreader*) must allow to combine back to the original form all the "pieces" composing the content.

A simpler and more robust solution concerns in users organizing themselves in server-based communities, exploiting data moving via the File Transfer Protocol (FTP) or by using ad-hoc file servers accessed via an IRC network. At the time of writing, the most popular networks of IRC servers are called *The Big Four*, and they are *EFnet*, *IRCnet*, *QuakeNet* and *Undernet*. Moreover, owing to its popularity, many IRC clients have been enhanced with an open protocol called Direct Client-to-Client (DCC) protocol, allowing two interfaces to directly communicate, preventing the need of having a centralized file server for routing the data. Even if limited and rough, the file-sharing practice exploited through IRC somehow resembles an ancestor of a modern file-sharing application.

Notice that all the aforementioned solutions imply the need of a well-suited centralized infrastructure, thus requiring sufficient availability (in the sense of the uptime performance)

[3]To be more precise, the most relevant bottleneck was actually imposed by the home users' access network.

and a non-negligible amount of bandwidth not to limit the sustainable client population. Such requirements, jointly with the low data-rate of early Internet accesses, limited both the amount of participants in file-sharing activities and the dimension of the exchanged volumes. The latter reflects in difficulties in exchanging multimedia material, being often in the order of tens to thousands of MBytes.

In this perspective, the need of exchanging files by using a dedicated solution for file-sharing was a **right** intuition for a **real** need.

1.5. Post-Napster File-Sharing

The fate of Napster inspired developers to primarily study more distributed solutions, making shutdown and resource tracking attempts more difficult. This effort spawned the second wave of file-sharing applications heavily relying on the p2p communication paradigm. As a consequence, modern systems are, more specifically, *p2p file-sharing* applications[4].

The current "new-wave" has been mainly spawned by five major systems. Namely:

1. **Gnutella**: originally developed by Justin Frankel and Tom Pepper, it relies on an unstructured p2p architecture with supernodes, which are a particular kind of node. Its overlay and major properties will be analyzed in Chapter 2. Even if still popular, it is not anymore in the first two top rank file-sharing applications, at least in terms of popularity across users. Yet, its development is well documented and the client interface has been ported over the majority of hardware and software platforms. In addition, many "proprietary" implementations (e.g., Limewire) also allow to mix different file-sharing applications in a unique software;

2. **eDonkey 2000**: developed by a software team called MetaMachine, it relied on a protocol called Multisource File Transfer Protocol (MFTP). The official eDonkey development has been stopped in late 2005. Also in this case, a legal struggle happened. Its major contribution is that it kickstarted the development of the **eMule** file-sharing application, that will be thoroughly analyzed in Chapter 6;

3. **Kazaa**: developed by Niklas Zennström, Janus Friis and Jaan Tallinn (the same trio responsible of the Skype[5] Internet telephony application, which also relies on a p2p-based infrastructure), it represented for a while the most widespread p2p file-sharing application. However, the presence of spyware, cycle-stealing software, as well as the menace of many legal threats, made Kazaa disappear. Nevertheless, Kazaa was the first application introducing interesting countermeasures to prevent traffic sniffing, such as random port hopping and flow cyphering among certain kinds of peers;

[4]This book covers this kind of programs, thus the definitions p2p file-sharing and file-sharing are assumed as interchangeable.

[5]A discussion about Skype is out of the scope of the book. However, we point out that the brand has been acquired for several billions of dollars and it offers also commercial services to millions of customers worldwide. Thus, people who want to gain more comprehension about business models and opportunities arising by the adoption of p2p technologies should investigate the success story of Skype. More details are available at `http://www.skype.com`.

4. **WinMX**: developed by a team called FastTrack, it does not anymore exist, at least, officially[6], but a small core of users continue to use a patched version to recover the lack of support from the original developers. WinMX was important due to its ability of handling both OpenNap-based networks and a distributed network similar to the Gnutella one. WinMX allowed to handle queues manually: even if it was considered as a degree of freedom, it was prone to user-abuse. As a consequence, many users migrated to file-sharing systems having an automated and more effective queue management, such as the aforementioned eMule;

5. **BitTorrent**: developed by Bram Cohen, nowadays, along with eMule, is one of the most adopted file-sharing platforms. Even if its evolution is fast-moving, also in the commercial field, it is basically a hybrid system with a centralized component, with a very simple design and very effective heuristics for managing the content distribution among peers. It will be extensively investigated in Chapter 7.

[6]Actually, it is not clear if a "cease and desist" has been issued to developers, but WinMX has been abandoned suddenly a couple of years ago, without any clear motivations.

Chapter 2

Peer-to-Peer Systems

This chapter offers an introduction to the p2p communication basics. Instead of presenting a portrait of the p2p paradigm "at large", it will focus only on such aspects that are functional to understand and engineer file-sharing applications. In fact, the majority of them relies on different flavors of p2p architectures. Currently, there is not a standardized nomenclature, and there is an open debate on defining a taxonomy for both file-sharing platforms and p2p applications in general. The key reason is rooted in the recent history of file-sharing: as discussed in Chapter 1, many successful systems have been spawned by brainstorms of very skilled users and developers, rather than by meticulous engineering and standardization processes.

The chapter deals with basic concepts such as the *churn*, that is the continuos process of nodes' arrivals and departures, the *overlay*, that is the virtual layout superimposed over the physical network and the different *p2p architectures* adopted for developing sophisticated file-sharing services. Since there is a wide acceptance only in classifying p2p file-sharing systems as *unstructured*, *structured* and *hybrid*, the chapter will solely focus on them. Hints about different properties (e.g., resistance against attacks) and specific benefits / drawbacks (e.g., when performing searches) are also presented.

2.1. From Client-Server to P2P

In a few words, the p2p model is the direct antagonist of the client-server communication paradigm. To better understand what p2p is, a vis-*á*-vis comparison with classical client-server architectures will be useful.

In client-server systems, there is a *unique* and *centralized* component responsible of running the service. In the very first client-server incarnations, all communications were assisted by the server and, without this centralized entity, the service would have been unavailable. In addition, clients connecting to the server could not communicate directly among one another.

Figure 2.1, on the left, depicts a classical client-server system. But, nowadays, many clients implement some mechanisms to establish a direct connection, even if the core functionalities are still demanded to the centralized component. For instance, this is the case of many IM platforms, which allow also to spawn point-to-point communications, e.g., for

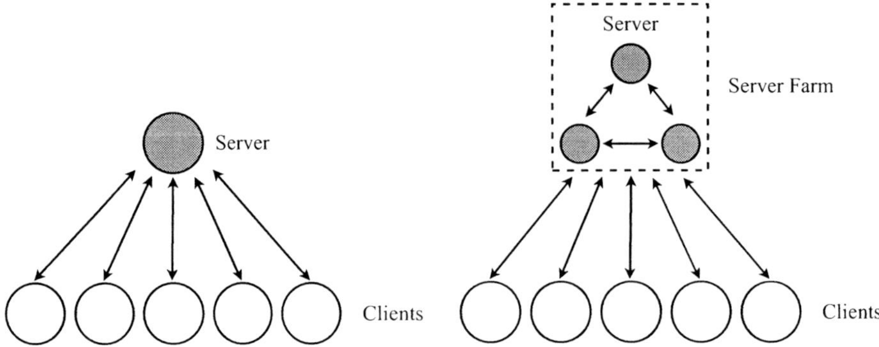

Figure 2.1. Different flavors of client-server architectures. On the *left* a client-server architecture with a single server, while, on the *right*, the service is delivered through server replicas.

exchanging files or performing audio/video communications.

Summing up, client-server frameworks are simple to understand, widely adopted and easy to set up. But they have some intrinsic limits and, most importantly, they exhibit a unique point of failure. As a result, malicious actions, such as Denial of Service (DoS) attacks, can paralyze the normal service providing phase. Some problems also arise with respect to performance and throughput bottlenecks: when a single centralized server cannot handle high client load, a common solution is to use a cluster of machines, allowing a higher transactional throughput. Then, replicating the servers is used to enhance the service capacity of a system. In Section 7.6., we will discuss how the service capacity is "automatically" handled in p2p file-sharing applications.

Figure 2.1, on the right, depicts a centralized architecture with a replicated server infrastructure. We point out that redundancy and machine replicas must not be confused with p2p architectures. In fact, even if highly distributed or replicated, the internetwork of the centralized components is always the unique entity responsible of maintaining the service up and running. In addition, different servers could interact with each other to deliver sophisticated services, but still relying on a fully centralized blueprint. Besides, a huge amount of centralized components cooperating to deliver a service could exhibit some behaviors of p2p systems, as it will briefly presented in Section 2.2.3..

2.2. The Concept of Overlay

One of the key characteristics of the p2p communication paradigm is its ability of building overlay networks. An *overlay network* is a network built on top of another one, as shown in Figure 2.2. Instead of "built on top", some engineers[1] do prefer to define an overlay as a network *juxtaposed* over another one.

In the case of p2p file-sharing applications, the overlay network is the one commonly exploited by TCP connections among different peers. In order to be effective, an overlay

[1] For instance, this is the case of the JuXTAposed (JXTA) project developers [6].

network must offer basic services, such as *addressing* and *routing*.

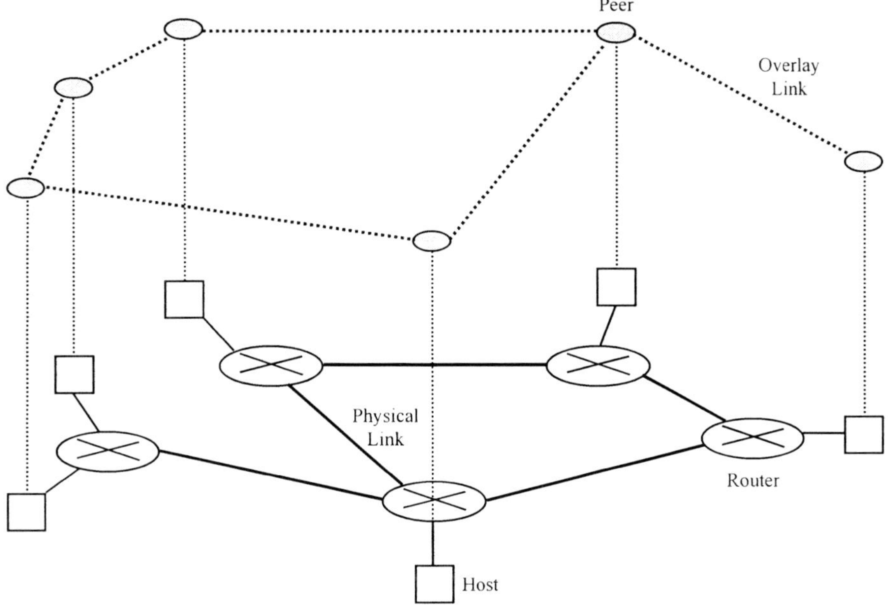

Figure 2.2. The overlay network composed by transport layer connections juxtaposed over the physical network deployment.

Another example of overlay could be the one implemented by the IP layer. In fact, IP offers addressing and routing strategies built on top of a physical infrastructure (that could be also highly heterogeneous). In another perspective, also the label switching mechanism implemented by the Multi Protocol Label Switching (MPLS) [7] technology could be perceived as an overlay.

Resorting to an overlay network to develop highly distributed applications offers isolation properties. In the case of file-sharing ones, the application logic is decoupled from the underlying physical network deployment, and it is also possible to use the overlay to overcome limitations imposed by the network itself. For instance, the overlay could be used to get over the lack of multicast support at the network level [8]. This is one of the most recent and active research trends in the field of overlay network utilization.

We point out the key reason under the choice of describing the three flavors of p2p networks (i.e., unstructured, structured and hybrid), and thus the resulting overlays. In fact, every p2p file-sharing application could be described as the superimposition of the proposed basic architectural blueprints. Therefore, analyzing the building blocks to compose complex overlays allows the reader to understand also future file-sharing applications, or those not covered in this book.

2.2.1. Overlay Topology Matching

As said, owing to its "independent" flavor, the overlay could be arranged separately from the underlying network infrastructure, in order to achieve given design goals. But, this decoupling may introduce inefficiencies in the exploitation of the available resources.

A toy example is provided in Figure 2.3, which depicts the underlying physical network and two possible overlays built on top.

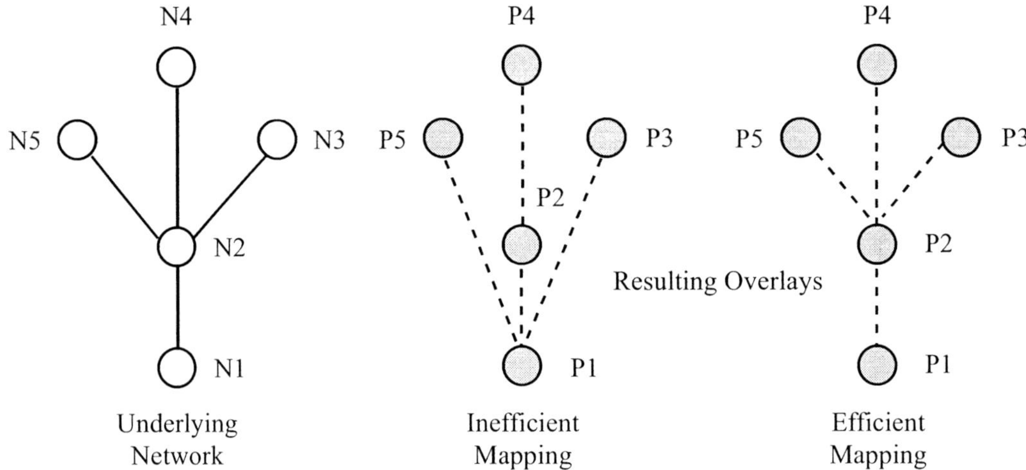

Figure 2.3. A toy example to show the concept of overlay matching. A physical network and two different matchings of the overlay network built on top are presented. *On the left:* the underlying physical network topology. *In the middle:* an inefficient overlay topology matching. *On the right:* an efficient overlay topology matching.

Let us define nodes N_i, with $i = 1,\ldots,5$ as the hosts connected to each other via the physical links $\overline{N_1N_2}$, $\overline{N_2N_3}$, $\overline{N_2N_4}$ and $\overline{N_2N_5}$. Let P_j, with $j = 1,\ldots,5$ be the j-th peer in the overlay corresponding to the i-th host. We define $P_j = f(i)$, where $f(i)$ is the mapping defined by the peculiar structure of the employed overlay.

Then, define also as an *inefficient overlay* the networking of peers interconnected via the following set of transport connections: $\overline{P_1P_5}$, $\overline{P_1P_2}$, $\overline{P_1P_3}$ and $\overline{P_2P_4}$. Moreover, define a path as the concatenation of two physical links and let us assume that the overlay operates the delivery of the data units in a store-and-forward fashion.

In this perspective, let us suppose that peer P_1 wants to send a message (e.g., a search query for a given file) to all the peers populating the overlay. Message to P_5 travels through the physical route $(\overline{N_1N_2},\overline{N_2N_5})$; message to P_2 travels through $\overline{N_1N_2}$; message to P_3 travels through $\overline{N_1N_3}$; P_2 forwards a copy of the message to P_4 through the link $\overline{N_2N_4}$. As a consequence of this particular arrangement in the overlay, the link $\overline{N_1N_2}$ must deliver the same (duplicated) message three times.

Conversely, let us consider as an *efficient overlay* the networking of peers interconnected via the following set of transport connections: $\overline{P_1P_2}$, $\overline{P_2P_5}$, $\overline{P_2P_4}$ and $\overline{P_2P_3}$. Let us evaluate the traffic produced in the underlying network to deliver the same message of the aforementioned example (i.e., a message from P_1 to all the peers). In this case, it is sufficient to send the message once to P_2, through the link $\overline{N_1N_2}$, and then rely on node P_2 to forward a replica to P_5, P_4 and P_3 through the links $\overline{N_2N_5}$, $\overline{N_2N_4}$ and $\overline{N_2N_3}$, respectively. Consequently, the link $\overline{N_1N_2}$ is traversed by the message once. This simple example permits to understand the importance of arranging nodes in the overlay in a rational manner to avoid a loss of efficiency.

As regards analyses of the mapping between a p2p file-sharing system and the underlying network, the pioneer work available in [9] analyzes the degree of mismatch between the overlay network of the Gnutella file-sharing system and the Internet. In order to reduce the needed effort, Ripeanu and Foster proposed a simple cluster algorithm. As a results, they conclude that the self-organizing network implemented by the Gnutella protocol does not match efficiently the underlying network infrastructure.

2.2.2. Metrics to Quantify the Matching

As explained in the previous section, quantifying the degree of matching of the overlay with respect to the physical network is a mandatory step to understand the impact of a p2p-based network application. In this vein, different metrics have been proposed, but a simple one allowing to quantify the efficiency of the matching is the *link stress* defined in [10] and [11].

Let us introduce its definition:

the *link stress* represents the total number of identical copies of a given data unit (e.g., an IP packet) transmitted over the same underlying physical link.

To complete such concept, [12] introduces the concept of *overlay cost*. Even if the study deals with overlays employed for multicast, it could be extended to other applications (i.e., p2p file-sharing, in our case) straightforwardly.

The *overlay cost* is defined as:

$$C = \sum_i S(i), \forall i$$

where i is a router-to-router link traversed by one or more overlay link, and $S(i)$ is the link stress experimented by the i-th underlying link.

The previous metrics emphasize the importance of having a good matching between the overlay and the underlying network. In fact, the reader could imagine the stress introduced by millions of users populating an overlay built for file-sharing purposes[2].

In order to better understand, let us calculate the introduced costs for the two toy scenarios depicted in Figure 2.3.

Let us start from the inefficient mapping scenario. The link stresses are:

$$S(\overline{N_1N_2}) = 3$$
$$S(\overline{N_2N_5}) = 1$$
$$S(\overline{N_2N_4}) = 1$$
$$S(\overline{N_2N_3}) = 1$$

[2]However, the overlay is commonly employed only to locate users and contents. As soon as the two endpoints are properly located, the file transfer is performed outside the overlay, e.g., by establishing a direct TCP connection. Thus, the resulting data flow is solely routed through the physical network according to the deployed L3 routing disciplines.

Then, the overlay cost for the inefficient mapping is given by:

$$C_{in} = \sum_{i=1}^{4} S(i) = 3+1+1+1 = 6$$

Conversely, let us consider the case when the efficient mapping happens. The link stresses are:

$$S(\overline{N_1N_2}) = 1$$
$$S(\overline{N_2N_5}) = 1$$
$$S(\overline{N_2N_4}) = 1$$
$$S(\overline{N_2N_3}) = 1$$

Then, the overlay cost for the efficient mapping is given by:

$$C_{ef} = \sum_{i=1}^{4} S(i) = 1+1+1+1 = 4$$

Thus, it is obvious that $C_{in} > C_{ef}$, i.e., an inefficient mapping stresses more the underlying physical network than in the case when an efficient one is employed.

For instance, let us consider again the toy example depicted in Figure 2.3. Imagine that the link $\overline{N_1N_2}$ is an intercontinental link: in this perspective, the cost (in terms of bandwidth and delay) of the stress introduced by the overlay becomes clear.

Thus, reducing the mismatch between the network and the overlay is important both for preserving the overlay characteristics and to not endanger the network infrastructure. Alas, due to the continuos process of nodes entering and leaving the system, thus mutating the overlay, the difficulties in precisely discovering the underlying topology and the huge computational effort required, it is very hard to compute a precise mapping for an overlay network. But, an engineering option is to avoid to optimize the matching in favor of employing a more sophisticated overlay and forwarding mechanism, as it will discussed in the following.

2.2.3. The Churn

While participating in a network service, users are free to enter and leave the system at any time in an independent and unpredictable manner. Due to the peculiarity of p2p environments, this phenomenon gains a particular importance.

Specifically, in file-sharing applications, a peer joins the overlay, it exploits the required operations and then leaves the system (suitable techniques to force peers to remain longer in the overlay to guarantee a proper degree of contribution to the system will be discussed in Chapter 5). The continuous process of peers joining and leaving the network is called *churn*[3]. As a consequence, the overlay mutates on a continuous basis, thus proper actions

[3]Here, with the term *churn* we intend the "overall" effect of nodes entering and leaving the network. However, several (concurrent) issues account for churn, besides the voluntary user disconnection. For instance, link outages and network congestions reflecting in expiring time-outs.

to maintain its consistence must be adopted. Quantifying the churn is then important to design an effective overlay, hence the file-sharing service. In addition, churn must be taken into account while conducting simulations of file-sharing applications to evaluate their performances and their impact over the network.

A valuable study about the characterization of churn in different file-sharing environment is available in [13]. Stutzbach and Rejaie discovered some interesting behaviors of the churn affecting file-sharing services. The major findings are: i) the dynamic of churns affecting different file-sharing services is quite similar, despite peculiarities of the different systems and ii) a large number of peers in the overlay exhibits a stable behavior, while the remaining turn over quickly. The first finding allows to make reasonable hypothesis when designing and simulating a p2p file-sharing application "from scratch". The second finding is an important propriety, still under investigation, allowing to define interesting disciplines to augment both scalability and stability of the overlay, e.g., by promoting some peers according to their uptime statistics. This could be exploited to dynamically promote peers to act as supernodes, which are particular peers as discussed in Section 2.3.2..

In the case of file-sharing applications, the churn has also another important effect. In fact peers abandoning the overlay also account for the increased/reduced availability of a given file(s). For this reason, many techniques to increase the file availability to cope with churn have been developed and they will be discussed in Chapter 5.

Lastly, to underline the importance of the churn, consider that it is not only a characteristic of p2p systems. For instance in [14], which analyzes some properties of a very large-scale cluster (i.e., the cluster of machines employed by Google to handle searches), it is shown that having a huge number of hosts with a reasonable low probability of outage reflects in a churn similar those populating p2p applications.

2.3. Unstructured p2p Networks

Unstructured p2p systems are frameworks that organize the peers implementing themselves without any enforcing algorithms or resorting to data structure with a global significance in the overlay. Thus, in unstructured p2p systems, the knowledge "stored" in the overlay is not sufficient to arrange peers in a more convenient way.

The unstructured flavor brings also some pros and cons. On one hand, unstructured p2p systems rely on simpler algorithms, for instance in terms of log-in and log-out operations. Besides, the operations devoted to the maintenance of the overlay only require to contacting peers in order to discover if they are active/inactive and maintain some states in a table. Such methodology is often defined as *ping-pong*, and it will be discussed in detail in Section 3.3.1.. On the other hand, locating peers and resources might be difficult. Due to the lack of a well-defined topological structure, the content is hard to locate and the system gives no hints about its possible location within the overlay. The most important consequences are that searches must be handled in a more "aggressive" manner, but to avoid to flood the network, they could be not exhaustive and subject to highly varying time delays.

Actually, there are at least two types of unstructured p2p networks. Even if it would be hard to classify unstructured p2p network, a widely accepted categorization is then introduced.

2.3.1. Pure Unstructured Systems

Pure unstructured p2p architectures are those where all the peers are equal standing and organize themselves in a completely independent manner. Thus, there are not "well-known nodes": functionalities are completely shared across the overlay network components. Figure 2.4, on the left, depicts the overlay implemented by a pure unstructured p2p system. The lack of a well-defined organization brings to major difficulties in developing algorithms devoted to search and route data. This characteristic introduces a problem related to information delocalization, leading to a situation where it is impossible to determine the peer that contains information of interest. Due to this fact, to perform searches (i.e., nodes,

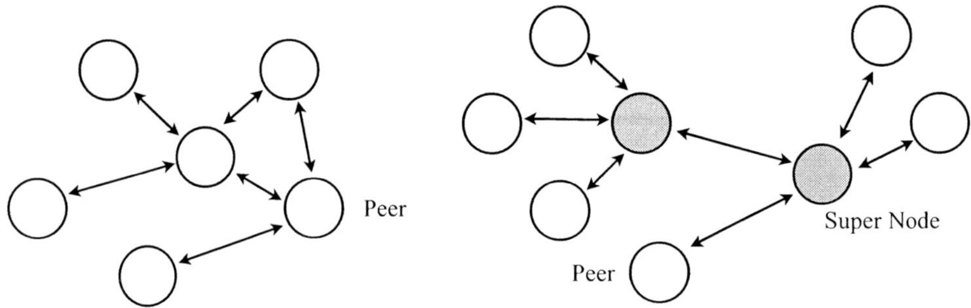

Figure 2.4. Generic snapshots of unstructured p2p overlay networks. On the *left* a pure unstructured p2p overlay network, while on the *right* a pure p2p overlay network with supernodes.

services and resources), many search protocols acting over a p2p unstructured overlay use a controlled broadcast algorithm, as explained in Section 2.3.4..

To reduce the localization indeterminacy, a widely adopted solution consists of leaving the pure p2p network architecture in advantage of a mediated p2p topology: some peers act as "mediation points", introducing a rudimental organization in the overlay. This brings to unstructured p2p systems with *supernodes* (often called also *superpeers* or *ultrapeers*, according to the specific file-sharing system).

2.3.2. Unstructured Systems with Supernodes

As already discussed, a typical solution consists in dropping the pure p2p network architecture by introducing some "special" peers. Often, the phrase "*some peers are more equal than others*" is employed to characterize this kind of overlay organization. Such nodes act as "mediation points", allowing a simplified network management: Figure 2.4 on the right depicts such situation.

The Gnutella network is one the first file-sharing applications relying on such organization, and another popular one was Kazaa. We will investigate other aspects of the overlay structure of Gnutella in the following of the book, but nothing concerning Kazaa. Therefore, being the latter already deeply investigated in literature, interested readers could find more details in [15] and references therein.

2.3.3. Unstructured Overlays and Self-Organized Networks

As a consequence of the freedom of peers in selecting interconnections to form the overlay (i.e., freely deciding the TCP connections to be established), unstructured p2p systems are able to produce *self-organized* p2p networks. The literature analyzed this peculiarity and demonstrated that the resulting overlay obeys a *power law* rule [16].

Specifically, the power law distribution closely matches the distribution of peers' connectivity that constitutes the overlay. Put briefly, such models state that an unstructured p2p system is composed by a huge number of peers with few connections and a small number of peers with a huge amount of connections. In more detail, the amount of peers with L links is proportional to L^{-k}, where k is a given constant, which depends on the network. Besides the availability of a model, power law networks (which is used here as a synonym of unstructured p2p networks) are also important for their properties, especially with respect to resistance against failures, churn effects or attacks.

Particularly, power law networks present a *high resistance against random failures*[4], *but they are endangered by well-planned attacks*. The resistance against random failures also explains their intrinsic attitude in handling deployments with high churn values, i.e., file-sharing systems.

A very interesting study about properties of unstructured overlays created by file-sharing applications has been conducted by Saroiu, Gummadi and Gribble, and it is available in [17]. By developing a crawler (this measurement methodology will be presented, in a general form, in Section 9.2.1.), Authors got a snapshot of a portion of the Gnutella network on February 16, 2001. Then, they studied the resulting overlays upon removing peers[5] both randomly and in a well-planned fashion. Such experiment allowed to highlight properties of the overlay network made by Gnutella client interfaces in the face of attacks or failures. This property is often defined as the *resilience* of the overlay.

Figure 2.5 depicts the different snapshots of the Gnutella network. In Figure 2.5(a) a Gnutella network composed by 1771 peers is depicted; notice that the overlay network is "dense" and well-connected. In Figure 2.5(b), the same network is presented, but after randomly removing a 30% of the peers composing the original network. Even if the network is composed by a smaller number of peers, still it is well-connected: there are no "isles", and all peers can, at least theoretically, reach all the others. Summing up: there are not any split spaces within the overlay. This is representative of the first part of the statement "*power law networks present a high resistance against random failures...*".

Lastly, Figure 2.5(c) depicts the same network of Figure 2.5(a) after removing only 4% of the peers; now rather than randomly, peers have been meticulously selected. Particularly, the peers removed were those characterized by a high amount of connections. As a results, the overlay is now composed by many isolated sub-networks, resulting in a fragmented architecture. This is representative of the last part of the statement "*...but they are endangered by well-planned attacks*".

This behavior is explained by the peculiar node distribution: maliciously removing

[4] A failure is representative of a node or link failure, malicious attack and the "natural" churn characterizing p2p file-sharing applications.

[5] Actually, instead of peer, Gnutella developers do prefer the term *servent*, that is the merge of *ser*ver and cli*ent*.

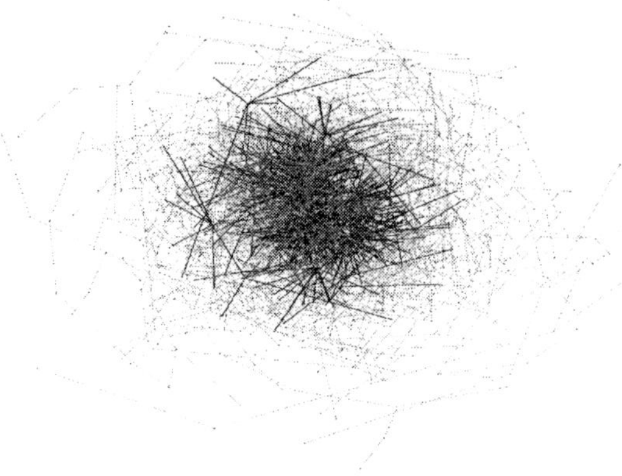

(a) A Gnutella network composed by 1771 peers.

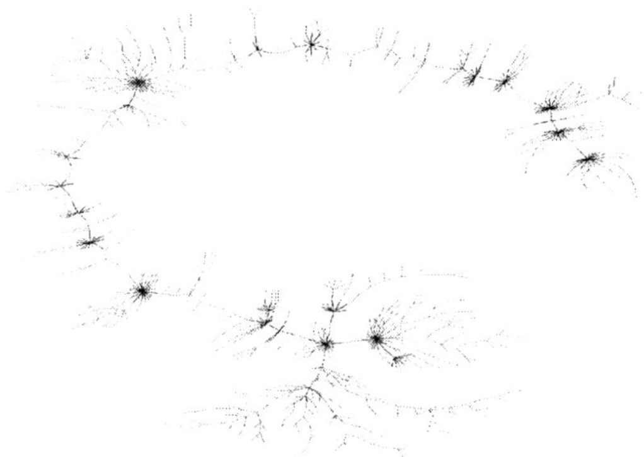

(b) The resulting Gnutella overlay after randomly removing 30% of the peers composing the original network

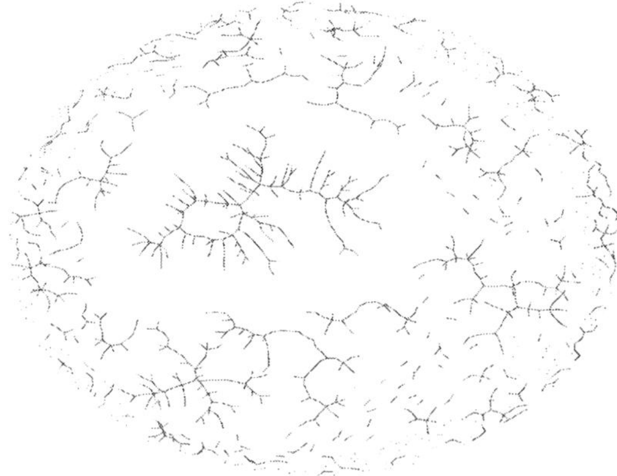

(c) The resulting overlay after removing only 4% of the peers meticulously selected.

Figure 2.5. Characteristics of the Gnutella overlay in the experiment conducted by Saroiu, Gummadi and Gribble. The presented plots are properties of the Authors. Used upon permission. The original work is available in [17]. Colors have been changed to improve readability for black and white printing.

peers with a high amount of connections will bring to a more fragmented scenario. To reduce the impact of malicious peers removal, the aforementioned Kazaa, which relied on supernodes, encrypted connections between two adjacent supernodes to increase robustness of the overall overlay.

2.3.4. Searches in Unstructured p2p Systems

In order to better explain how the overlay organization can interfere with routing and search strategies, we present an example of how a search is typically handled, in unstructured p2p overlay. The scenario here introduced represents a search query in a Gnutella-like network. Specifically, for didactical purposes, we refer to version 0.4 of the protocol specification, since from version 0.6 of the specification a different overlay organization, with a Time To Live (TTL) limited to 2, has been employed. In this example, queries are broadcasted over the network and to avoid an uncontrollable load, a proper limiting mechanism must be introduced. Gnutella uses a very common approach, that is introducing a special counter, called the TTL. Queries are generated and then associated to a an initial value of TTL. Then, the TTL is decreased by each peer that forwards the query. When the counter reaches the 0 value (it expires), the query is not forwarded anymore. In the official Gnutella implementation, the TTL limit is set to 7.

As explained earlier, in unstructured systems peers' organization is not enforced and peers organize themselves in a random manner. Thus, it is impossible to known *a-priori* where the desired information is stored, as well as having hints about the portion of the network where the desired information is stored with higher probability. Actually, for this reason, structured p2p has been developed: to overcome this *organized anarchy* behavior.

For the sake of simplicity, let us assume that the overlay is structured like a tree, i.e., no cyclic paths are present. This hypothesis does not prevent to examine searches in a general way, since, as it will be explained later, it is not too restrictive. Figure 2.6, on the left, depicts the reference scenario. Let us suppose that the grey peer, called the *Query Originator* wants to search for a content, say a multimedia file. Peers filled with the black color are those containing the required contents.

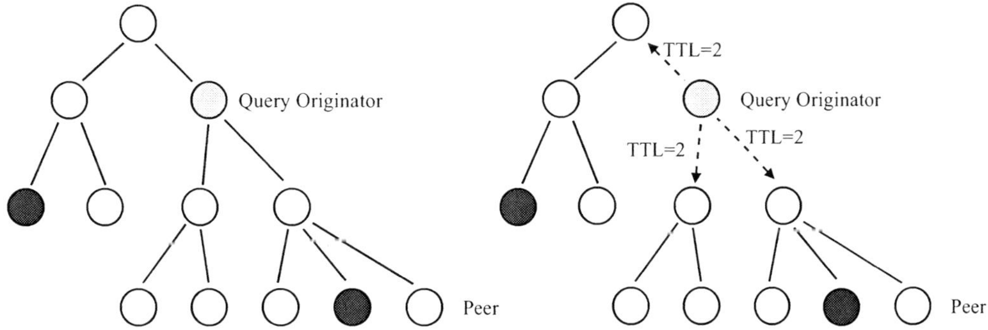

Figure 2.6. *On the left*: the reference toy scenario. For the sake of simplicity, the overlay network has been assumed with a tree topology. *On the right*: first step of a distributed search in an unstructured p2p network. The initiator is the peer filled with a grey color and with the boldface border, while black filled peers own the desired content.

As a first step, depicted in Figure 2.6, on the right, the originator can only send a query to its neighbors, being only aware of them. The query has been associated with a TTL=2. Clearly, this value of TTL does not assure that the desired content will be found, and usually it is a tradeoff among three quantities: *speed*, *generated load* and *accuracy*. The adjacent peers analyze the content of the query. If they own the desired content, they send back a *query hit response*. Then, they decrease the TTL and, if it does not expire, the original query is forwarded to their neighbors. Figure 2.7 depicts two iterations of the process.

When a peer owning the content specified by the query has been reached, it generates a query hit response, and the result is routed back to the query originator.

In order to actually transfer the desired content, a direct communication will be established by using a proper protocol, i.e., a proper signaling mechanism. In this case, the Gnutella overlay is not employed for data delivery, i.e., it does not convey nor route data related to file transfers.

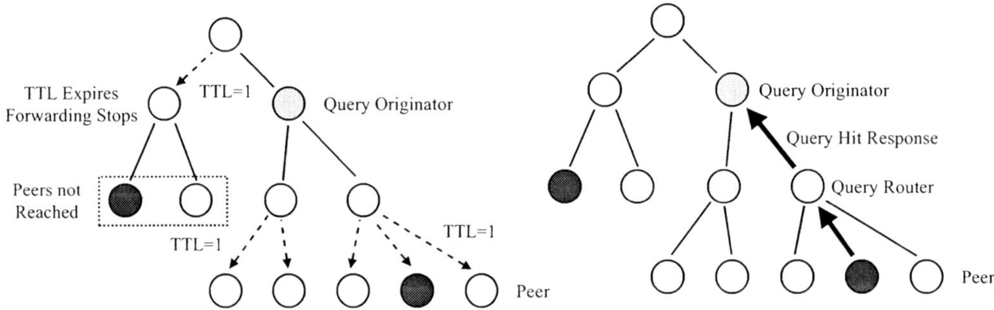

Figure 2.7. Evolution of a distributed query in an unstructured p2p network. The iterations for $TTL = 2$ and $TTL = 3$ are depicted.

Obviously, this mechanism also relies on state information within peers. State information is needed to route back the query and to avoid loops (this is why we can, at least approximatively, consider the overlay as a tree when routing queries). To cope with such requirements, Gnutella adopts Unique IDentifiers (UIDs) to recognize peers and their queries. A peer responsible of routing back the result(s) is informally called *Query Router*. The explained approach is known in the literature as *flooded request model* [18] since at each iteration, the overlay is flooded: each peer contacts all the available neighbors. Hence, this method suffers of poor scalability: for large "communities" of peers, the amount of traffic generated by the system will increase explosively, as well as the state information needed in each peer. Nevertheless, as discussed in Section 2.2.1. the usage of this mechanism with a non-optimal overlay matching, jointly with the churn affecting the system, could bring to high inefficiencies. For such reasons, the scalability of this method is questionable.

When deployed in a large scale system or, more generally, in overlays having a size not comparable with the selected value of the TTL, this algorithm has also another drawback: searches are not exhaustive. In fact, it could happen that a content is not localized because it is stored within peers out of range of the given TTL value (see Figure 2.7 on the left, for an example).

In order to better understand the amount of traffic generated by flooded request model-like queries, we briefly introduce a simple formula based upon a simplified Gnutella-like model. The formula has been derived under the following assumptions: i) the overlay has a tree structure; ii) queries are routed by using a flooding algorithm; iii) every peer is connected in the overlay to other n different peers; iv) data to be searched is spread by obeying to a random distribution.

Then, the number of contacted peers, $N_{Peers}(TTL)$, which is function of the maximum value of the TTL, is given by:

$$N_{Peers}(TTL) = \sum_{i=1}^{TTL} n(n-1)^{i-1}$$

Notice that the TTL value accounts for the exponential growth of the resulting number of dispatched messages.

The presented search mechanism has also some impacts over the user experience of the file-sharing application.

The main considerations follow:

- *users receive search results with delays*: since query hits are delivered from different parts of the network and by peers having different availability of resources, results may not appear in the client-interface in a "one shot" fashion. Conversely, according to the status of the overlay, a user may wait for a considerable amount of time before receiving the results;

- *the perceived accuracy may decrease*: due to latencies introduced in routing back query hits, the results received may loose accuracy. Particularly, the churn may reflect in some results not delivered back to the Query Originator (i.e., a Query Router goes offline while routing a hit). Nevertheless, some query hits may arrive after some internal time-outs expire within the client interface, as to avoid waiting for completion for an unbounded time frame;

- *this strategy may trigger additional churn due to users' intervention*: since TTL accounts for non exhaustive searches, a user may compulsively log-out and log-in *ad libitum* in order to change its location in the overlay trying to reach unavailable contents due to the distance in the overlay. Such problem will be investigated, in more detail, when discussing the *first-node* problem in Section 3.2.2..

2.4. Hybrid p2p Systems

This kind of systems are also commonly called "centralized" systems, as to emphasize the presence of a central component exploiting some duties. This flavor of p2p networks is somehow the superposition of two different architectural blueprints, specifically, centralized systems and p2p architectures. In this perspective, we do prefer the nomenclature *hybrid p2p* network, and we will adopt it through the rest of the book.

Then, some duties are performed by the p2p architecture, while other tasks are typically carried out via a client - server architecture, as depicted in Figure 2.8. We point out, that centralized components may also interact among each other. Other kinds of hybrid architecture are also those employing different p2p flavors mixed at once.

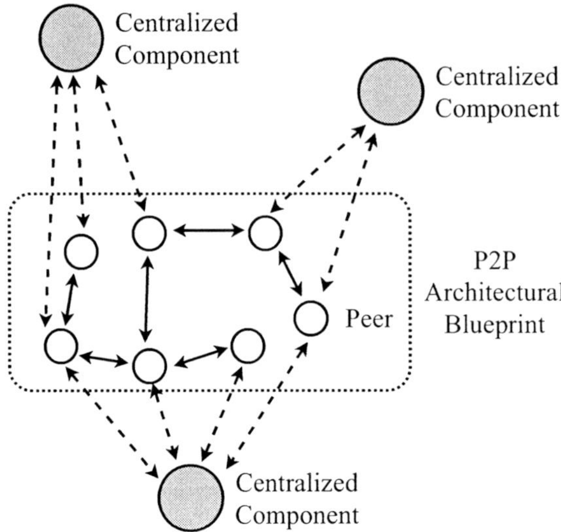

Figure 2.8. Snapshot of a hybrid p2p system. Client-server relationships are depicted with a dashed line, while p2p interactions are depicted with solid lines.

In order to clarify this concept, we introduce the Napster architecture both as an example and for historical reasons.

2.4.1. The Napster Architecture

As said in Chapter 1, Napster relied on a centralized component, called the Napster Server. It was adopted for logging-in/logging-out operations and to perform content searches. As soon as a client interface logged-in, it uploads the list of shared files to the Napster Server. Then, if a user performed some searches, the user would issue a request to the centralized component and wait for a list composed by the IP addresses of the nodes having the requested resources. This approach was easy to implement, quick (being the searches locally handled) and exhaustive (being the queries processed by a component containing the network's complete knowledge). However, it represented the main flaw of the system: without the Napster Server, users were unable to access the service. Besides, users already in the system were not able anymore to perform searches and to locate other peers. Figure 2.9 depicts the architectural blueprint of the Napster service.

Differently from the classical FTP data transfer, files were directly exchanged between two clients, thus the need of using a protocol for managing the direct interaction. The centralized component was then employed for orchestrating the entire service, hence resulting in a critical component. We point out again, that this caveat was exploited by lawsuits and the system was shutdown.

For the sake of completeness, we briefly describe some internals of the reverse engineered implementation of the Napster Client-Server Protocol, in order to underline its simplicity and its similarities with the IRC one. Napster adopted the TCP for client to server communications. The server usually accepted connection on ports 8888 and 7777. Mes-

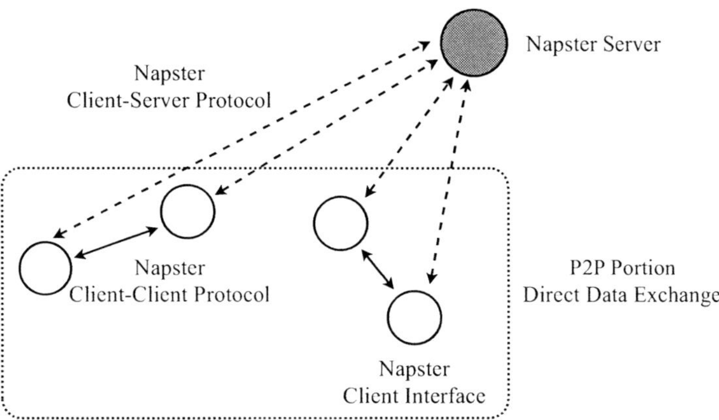

Figure 2.9. The Napster architectural blueprint. Notice client interfaces interacting with the Napster server farm via the Napster Client-Server Protocol (depicted as a dashed line), while the client-to-client interaction is done via the Napster Client-Client Protocol (depicted as a solid line).

sages exchanged among the two endpoints had the following format: `[length] [type] [data]`. The field `[length]` was 2 bytes long and contained the amount of data embedded in the `[data]` field. Notice that this field was used to properly parse the embodied messages and it differed from the philosophy of the IRC protocol of limiting a message with the `/r/n` trailing sequence. The `[data]` field contained messages formatted in plain ASCII strings. Lastly, `[type]` was also 2 bytes long, and it specified the flavor of the message contained in the Protocol Data Unit (PDU).

Clearly, this protocol is easy to implement, for instance, by stuffing a well-formatted ASCII string into a standard `socket( )` ANSI-C system call.

2.4.2. The OpenNap Architecture

As shown in Figure 2.9, Napster relied on two different proprietary protocols, that have been reverse engineered and documented in [19]. Actually, an open source implementation of the Napster protocol is available and named OpenNap [20], but it appears that Napster-based technologies are pretty outdated and no longer adopted, at least by the masses.

Like the original Napster, OpenNap relies on servers, but many servers can be used simultaneously and linked together to form a so called *server-net*. Figure 2.10 depicts a file-sharing service exploiting OpenNap servers. This architectural evolution has been done for the following reasons: i) since servers are not hosted by professional organizations, the distribution of the load rises the availability of the service and reduces the resource requests in terms of CPU, memory and bandwidth; ii) distributing the critical functionalities over different remote users dramatically reduces shutdown risks.

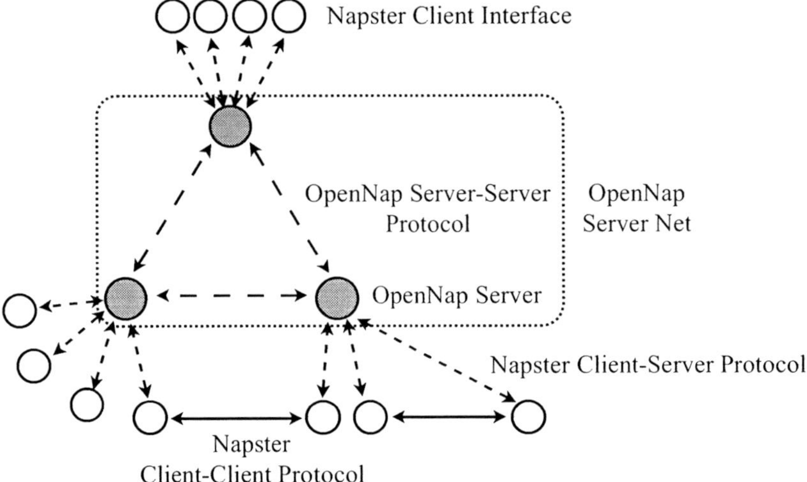

Figure 2.10. A file-sharing service exploiting OpenNap servers. Different independent servers are arranged and linked together to form an OpenNap server net. Client interfaces exploit the server net to page users. Notice that the server net is perceived by clients as a monolithic entity.

2.5. Structured p2p Systems

In *structured p2p* systems, the peer organization is enforced by some well-defined policies. Then, the topological characteristics of the resulting overlay are predictable and suitable for developing efficient search and routing strategies. However, these characteristics do not come for free: log-in and log-out operations are more complex, as well as state information in every peer. Structured p2p systems rely on a special distributed data structure called Distributed Hash Table (DHT).

DHTs, like a normal hash table, allows to store a (*key*, *value*) pair, but in a distributed way. In other words, a peer is responsible of a given (*key*, *value*) pair; such a node is said to be the responsible for such a *key*.

Different flavors of structured p2p systems exist, but the most adopted ones are those that produce a ring in the overlay. For the sake of completeness, we will briefly discuss Chord, that is a ring-based structured overlay for look-up operations. In Chapter 6 we will describe the XOR-based Kademlia overlay, which is at the basis of the Kad network used in the eMule file-sharing platform.

The Chord [21] protocol implements just one operation: given a key, it maps the key onto a node. Chord uses a variant of consistent hashing [22], [23] to assign keys to Chord nodes. Consistent hashing has also load balancing properties, since each node receives roughly the same number of keys, and it involves relatively little movement of keys when nodes join and leave the system. Each Chord node needs "routing" information about only a few other nodes. Being the routing table is distributed, a node resolves the hash function by communicating with a few other nodes. Chord maintains its routing information as nodes join and leave the system. Since, in p2p file-sharing applications, nodes' arrivals and departures are unpredictable (especially in large scale systems), Chord has also algorithms for maintaining this information in a dynamic environment.

Owing to its design, a very small amount of routing information suffices to implement consistent hashing in a distributed environment. Each node needs only to be aware of its successor node on the ring. Queries for a given identifier can be passed around the ring via these successor pointers, until they first encounter a node that is a successor of the identifier. However, this strategy is non-optimal, since it may require traversing all the N nodes composing the overlay, to find the requested mapping. Figure 2.11 depicts two different look-up flavors: the bold line represents a non-scalable look-up, e.g., the number of nodes to contact increases with the overall peers population; instead, the finger table allows the look-up procedure to scale.

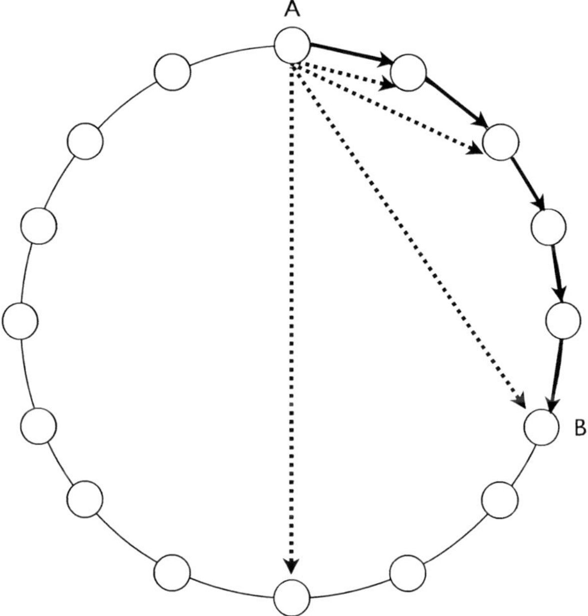

Figure 2.11. Peers organized in a Chord ring. Without any accelerations (i.e., fingers according to Chord nomenclature) the look-up from **A** to **B** has low performance: all the intermediate peers must be traversed (depicted with a solid bold line). When employing fingers (depicted with dotted lines) looking-up in the ring is accelerated. Notice that fingers used here have not been calculated precisely. Instead, they have been depicted only for didactical purposes.

Let us consider the non-scalable look-up: node A wants to find a route to node B; without any improvements, the messages are routed among all A's successors until B is encountered.

In order to speed-up searches and to dramatically limit the maximum number of needed entries, each Chord peer maintains a special routing table called *finger table*. In Figure 2.11, the dotted path depicts the shortcuts provided via the finger table (paths are not precisely computed, instead they are provided as a simple reference). Specifically, each peer maintains a routing table of (at most) m entries, with $m \ll N$, where N is the total number of peers composing the ring. The i-th entry in the table points a node j that succeds at least by 2^{i-1} nodes. As presented in [21], in a network composed by N nodes, with a base b the amount of entries maintained by each peer obeys the rule: $(b-1)\log_b N$, thus reflecting in a $O(logN)$ complexity.

Peers can continuously enter-leave the system in an unpredictable flavor, thus the resulting churn can alter (also in a significant way) the composition of the Chord ring. However, peers leaving the systems can take the overlay to an inconsistent state, due to incomplete (or erratic) log-out procedures. As a consequence, fingers must be constantly updated to reflect changes in the ring. To cope with this, Chord relies on a procedure called "stabilization". Every T_{stab} amount of time (that can be specified) peers are forced to recalculate their entries, hence removing or updating invalid fingers. Stabilization will be discussed in Section 3.3.3..

Chapter 3

Anatomy of a File-Sharing Service

This chapter thoroughly analyzes the different functionalities composing a full featured file-sharing service and their locations within the overall distributed architecture. The detailed breakdown is provided in order to better understand where the usage of p2p solutions (presented in the previous chapter) is more convenient.

Put briefly, the chapter discusses the basic building blocks needed to implement a successful and modern p2p file-sharing platform. Nevertheless, standard additional features populating modern client interfaces will be discussed.

3.1. The Main Functionalities

In order to properly work, a file-sharing system infrastructure must perform several basic operations. The latter could be located within a centralized entity, in a distributed way within client interfaces or in a mix of the two flavors.

The needed features are summarized as follows:

- network kickstart[1]: a user must be able to enter the overlay and join in the file-sharing service. Such operation could be trivial, for instance when a well defined centralized component is present, or could be more complex according to the degree of distribution of the overall service, and the number of overlays adopted;

- network management: the overlay must be kept in a consistent state, by periodically checking its configuration (i.e., routing tables, peers population, ...), and by making the needed adjustments. Network management could be performed by using both centralized or distributed heuristics;

- search handling: the file-sharing infrastructure must be able to handle users' queries both in terms of file searches and users location within the overlay. Again, such operations could be exploited either in a centralized or in a distributed fashion;

- file-exchange disciplines: such mechanisms are those responsible of effectively transferring the data composing the file from one peer to another;

[1] Often, network kickstart is also defined as network bootstrap.

- enforcing heuristics: since users are also responsible of delivering contents and of actively exploiting duties for maintaining the infrastructure operative, some enforcing heuristics are often needed and employed. For instance, peers could not guarantee a proper degree of contribution, then suitable enforcing mechanisms must be in place. A detailed discussion about the most popular heuristics will be introduced in Chapter 5, while here we only present where they are usually located, with respect to the overall system architecture, and what kind of information could be exploited;

- assisting communication among peers: some peers could not be able to contact other nodes due to particular network configurations or policies deployed in the machine running the client interface (e.g., a personal firewall blocks inbound traffic or selectively denies access to specific ports). In this perspective, some features to help peers in participating the system are often employed;

- companion services: companion services are such functionalities not strictly related to file-sharing operations, but are deployed to augment the attractiveness of the system, i.e., IM services.

3.2. Kickstart Methods

Due to the presence of churn, accounting for the high degree of mutability of the overlay, and the presence of hosts accessing the Internet by relying upon dynamically assigned IP addresses, a node who wants to join the file-sharing network must obtain a prior knowledge about available service entry points to the overlay.

Roughly, this could represent a problem related to managing the log-in of new nodes within a preexistent system. In the p2p file-sharing jargon, methodologies exploited to build the overlay and to assist nodes to enter are often called *kickstart methods*.

Several solutions exist to cope with the aforementioned issues: anyway, almost the totality relies on the presence of a well-known centralized component. In this perspective, the highly autonomic and distributed nature of p2p represents here one of its major drawbacks. Needles to say, if the centralized component devoted to inform new peers about a possible entry point in the network fails, new nodes will be impeded to join the overlay.

As explained, there are no standardized architectures or procedures to allow a new peer to become conscious of where to enter the overlay. At the time of writing, the deployment of an ad-hoc centralized directory is the preferred one.

Before investigating the most popular methodologies, we briefly report less popular ones, for the sake of completeness.

- Manually updating a list of well-known hosts: this approach resembles the *host.txt* file maintained in the early days of the Internet, for resolving names prior to the invention and the deployment of the Domain Name System (DNS). One or more files containing a list of on-line peers have to be made available for download. Such lists are typically delivered via web-pages, or by client interfaces via automatic download procedures. Even if this was feasible in ancient systems populated by a small subset of nodes with a barely long lifetime, it is nowadays abandoned due to its scalability problems. But, for laboratory trials and small testbeds, this solution is still valid[2]. Nevertheless, it is also possible to handle servers manually, being typically only a small number of entities, if compared with the overall peer population (as explained in Section 3.2.5.).

- Hardcoding some well-known addresses into the client-interface: the IP address (or the DNS name) is hardcoded in the client interface. This approach was used in Napster, and it is no longer employed, since, if the centralized component is shut-down, due to the flaws introduced by the hard-coding methodology, it becomes complex to find some workaround[3].

- Peer Caches: peer caches should not be confused with centralized Peer Caches (when needed, to avoid confusion, we refer to the latter as WinMX Peer Caches). The concept of peer cache has been introduced in the Gnutella system and relied on each peer maintaining a set of known IP addresses belonging to other known peers. Then, before requiring assistance from other components dedicated to bootstrapping, a peer tries to contact known peers to see if they are still in the overlay and can handle the operation needed to join the file-sharing service.

- Using the reverse signaling flow from previous accessed overlays: this technique is mainly adopted for file-sharing systems and it is just a proof-of-concept, being very daring and successful with a very low probability. However, it explains part of the soul of p2p, and hence it is worth mentioning. Typically, in file-sharing systems, owing to the high number of users and the limited bandwidth available for serving them, users are remotely queued. As soon as the queue is exhausted, the remote client calls back the original peer, informing that it can go ahead with the download operations. Let us suppose that the original peer owing the address IP_A has logged off, and a new host, which wants to join the p2p system has gained the address (e.g., re-assigned via a DHCP server). It could happen that the remote peer performs a callback to IP_A and, if the new host is interested (and the client interface has a proper logic to exploit this "unsolicited reverse signaling"), it can retrieve the address of that host and use it as entry point for the p2p system. Clearly, this approach is quite chaotic and without any assurance; hence, it has never been used outside lab tests.

- Using Zeroconf-like strategies: Zero Configuration Networking [26] is an IETF Working Group, which developed techniques for finding services without a directory server, by using different techniques, like the multicast DNS and the DNS Service Directory. Zero Configuration protocols are available for developers (e.g., on MacOS X via the Bonjour technology) and are also suitable for kickstarting network services without any well-known host. Typically, Zeroconf works only on local networks and, hence, it is seldom adopted for deploying Internet-oriented p2p applications.

Notice that the aforementioned methods and the one that will be discussed in more detail in the following, could be mixed and used jointly.

[2]For instance, this solution has been successfully employed for building an overlay to support QoS in WLANs, as explained in [24]. In addition, such a method allows to quickly have a prototype also being able to communicate with an emulated environment [25].

[3]A typical and simple workaround could be similar to the one described in Section 3.2.2..

3.2.1. The First Node Problem

As hinted, due to their distributed flavor, p2p systems present many problems when a new user wants to join in. More generally, p2p suffers many drawbacks in bootstrapping the network. The two major issues are: i) due to churn, it is not sufficient to cache peers available in the system at a given time stage; ii) since file-sharing systems are often user-operated, the diffusion of dynamic IP addressing schema prevents the possibility of relying on known peers from a session to another.

The category of problems to be solved to join an overlay is commonly called the *first node problem*. Basically, a new peer who wants to join the network must know a priori at least one node to be able to join the overlay.

Obviously, this issue is not present in a centralized architecture, since it is sufficient to know the server's address, that is often resolved via DNS mechanisms (e.g., for web-based service, `http://www.apple.com`), or in presence of a well-known fixed addressing scheme. This mechanism shifts the problem in finding a proper bootstrap entity into a pointer to such a entity.

3.2.2. WinMX Peer Caches

We present *WinMX Peer Caches* that are very close to bootsrap servers. We refer to them, in the rest of this section, simply as Peer Caches. We decided to present this network kick-start method because it is general enough to cover and explain the majority of centralized methods employed to solve the first-node problem. Figure 3.1, on the left depicts a peer that wants to join a generic p2p network; in order to enter, it must become aware of some entry point. A caching infrastructure is something in the middle between a centralized component and an automatically managed list of well-known hosts. This allows to become aware in an automatic way of a sub-group of hosts participating in the p2p network. Usually, the cache infrastructure will not assist the peer in logging-in, but it will only provide some entry points to the overlay network.

In order to discover some possible entry points, a node polls the Peer Cache by sending a `Peer List REQ` message, thus asking for a list of well-known peers. Notice that at least the location of the Peer Cache, in term of an entry in the DNS, or its IP address must be known. Then, the Peer Cache will provide some IP addresses of peers composing the overlay. In Figure 3.1 the requesting peer obtains the IP address 192.168.44.15 of a peer in the overlay as a possible entry point. At this stage, the peer tries to contact the host with the address 192.168.44.15, e.g., by establishing a direct TCP connection and then by using some handshaking procedure. If the attempt to join the system fails, it will ask the Peer Cache again for another address. Actually, to avoid latencies and waste of resources, a Peer Cache can provide to a requesting node a block of up to N_{Ep} addresses at once, where N_{Ep} is a parameter to be selected while designing the system. Upon successfully joining the overlay, the peer will become itself an entry point for future log-in attempts. Then, it will ask the Peer Cache to update its content, by sending a `List Update` message, as depicted in Figure 3.1 on the right. The protocol employed by WinMX is proprietary; regarding a possible unified procedure for using Peer Caches, in [27] and [28] an infrastructure based on the Session Initiation Protocol (SIP) standard protocol is presented.

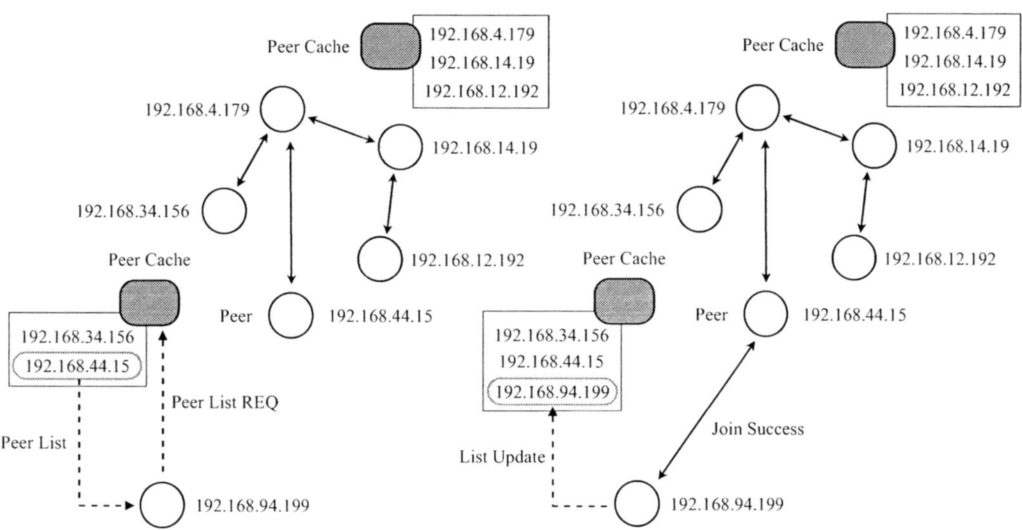

Figure 3.1. Snapshots of a system employing Peer Caches to handle the bootstrap process. *On the left*: a node querying the Peer Cache to find possible entry points in the overlay. *On the right*: the peer has successfully joined the overlay, and it informs the Peer Cache to update the state of the network accordingly. Notice that, in this example, private IP addresses have been used intentionally.

Concerning the possible entry points to the system, also the specific flavor of the overlay plays a role. The proposed approach has been successfully exploited for the WinMX overlay, which implements an unstructured system with supernodes (explained in Section 2.3.2.). In fact, as a prerequisite, in WinMX, supernodes are nodes able to accept inbound connection requests, thus being able to receive and handle join requests. For this reason, the presented mechanism is quite efficient since Peer Caches store and deliver information only about supernodes, which are a small subset of the overall peer population.

Always referring to Figure 3.1, notice that there are multiple Peer Caches. In fact, to provide a more scalable and fault-tolerant infrastructure, multiple Peer Caches could be adopted. In the case of WinMX, different caches were deployed for managing the kick-start of a node belonging to a given zone in the world (e.g., Europe, Asia, America, ...). In addition, caches were maintained by the software developers and their addresses were hard-coded within the client interface. Such characteristic was exploited by WinMX developers to do a kind of euthanasia to the file-sharing network when experiencing legal pressures. As mentioned in Section 1.5., users recovered this by setting up homemade peer cache and forcing the client-interface to search peers there, by simple overruling the hardcoded URL, e.g., by adding an entry in the *host.conf* file of the operating system.

Lastly, as discussed in Section 2.3.4. searches in file-sharing systems based on unstructured p2p could vary their effectiveness (e.g., the amount of results) according to the location of the peer in the overlay. This is due to the limited horizon enforced by a finite value of the TTL. Then, users may decide to log-in and log-out in order to obtain another subset of IP addresses to join the overlay connected through other peers, then augmenting the chance of finding useful results. This behavior, if repeatedly performed by a vast amount of users simultaneously, could generate an excessive load of requests to the Peer Caches.

3.2.3. Possible Pitfalls of Centralized Peer Caches

Being a centralized component, Peer Caches suffer of the same pitfalls of other similar solutions. We point out that they solely store information about how to reach a subset of nodes participating in the overlay, rather than file lists or contents. Thus, they partially relieve their maintainers from legal responsibilities and they are also less critical than a fully centralized solution. For instance, if Peer Caches depicted in Figure 3.1 stop working, new peers are prevented from joining the system, but without endangering the already established overlay.

In addition, developers, which hardcode the Peer Caches IPs or URLs, within the distributed software, put an hazard that can be effectively exploited to prevent access to the system, for instance by ISPs and network administrators.

A possible way to force malfunctioning when in presence of such setting is by using the *the poisoned DNS* technique. Actually, such technique could be used for blocking all kinds of URLs, but it was successfully employed to block access to file-sharing services, instead of blocking a wide range of TCP/UDP ports. The poisoned DNS consists in analyzing and filtering DNS traffic and requests. Its key assumption is that application developers do not "hardcode" the IP address of a "kickstarter" entity inside the application, but they rely on DNS entries (e.g., `bootstrap.example.org`) to have more flexibility.

The poisoned DNS approach exploits invalid DNS information sent back to the unsuspecting user. To exploit this technique at least two DNS servers are needed: one that is under control, providing invalid IP addresses, and another one that will resolve valid DNS queries [29].

3.2.4. Relying on a Fully-Centralized Component in the System

With *fully-centralized* component, we define an architectural entity that centrally manages all the duties needed to guarantee the file-sharing service to work properly. Thus, it is responsible of the log-in and log-out procedures (i.e., also to perform kickstart), but also of supporting search operations. Summing up, it represents the "intelligence" of the overall p2p file-sharing infrastructure, while the p2p portion is simply employed to distribute the traffic load generated by the exchange of files. For instance, this approach has been used by Napster and by many other systems before lawsuits "invaded" the p2p world. The centralized component, on one hand, allows to bootstrap a system in a simple way, while, on the other hand, it allows the system to be simply shut down. In this perspective, this approach has been progressively abandoned in favor of more distributed ones. The major benefit of WinMX Peer Caches is that allowed to decouple the bootstrap duties from the other ones (e.g., searches). However, modern systems, being aimed at the delivery of legal material, are proposing again the centralized-oriented solution.

In this perspective, the BitTorrent's tracker implements this solution; in this case, the first node problem becomes the same as finding the proper URL for a repository of meta-information, like `http://www.dimeadozen.org`.

3.2.5. Using Lists of Servers

Servers are employed by peers to join the overlay. In order to keep the network in a consistent state, servers are interconnected one to another, as to compose a network (sometimes called a *server net*). This could be assumed as an extension of the previous approach, but more robust. In fact, servers are ad-hoc machines supposed to have a high uptime. While having a list of active nodes could be inefficient due to the autonomic nature of hosts and for effects due to churn, server population is supposed to vary more slowly. Anyway, this approach is still adopted (for instance in eMule, which will be extensively investigated in Chapter 6), but the p2p infrastructure must be designed to support interaction with a server facility.

3.2.6. Knowledge Percolation from Already Accessed Overlays

Sophisticated p2p file-sharing applications often employ different overlays simultaneously, in order to locate a huge amount of content in a faster and efficient manner. For instance, this is the case of the eMule application, where at least two different overlay networks exist at once: the server-centric eD2k infrastructure, and the Kad serverless network. As explained, relying on a centralized component accounts for an easy solution of the first node problem. Then, as soon as a client enters the eD2k network, it could ask to peers for possible entry points for the Kad network.

3.3. Network Management

Actually, the term "network management" is quite vague and broad, but here it is used to define operations devoted to maintaining the overall infrastructure working properly. The most important task is to keep the overlay in a consistent status, for instance to handle the churn. In Chapter 2, different form of overlays and architectures have been discussed. In this perspective, the management is the task responsible of creating the explained environments in order to provide a suitable playground for performing file-sharing operations.

3.3.1. Ping-Pong Methods

Ping-pong is a distributed overlay management method resident in client interfaces mainly adopted in the Gnutella network and in hybrid systems. Moreover, it is also employed in structured system to recognize whether or not an entry in the finger table is alive. It is often utilized along with stabilization algorithms, so it will be discussed separately in Section 3.3.3.. Peers exchange short messages in order to explore the network. In addition, it is also useful to discover and update connectivity status, thus mutating the status of their neighborhood.

Basically, two messages called `PING` and `PONG`, i.e., a kind of probing request and response, respectively, are exchanged. However, this method lacks of scalability, in the sense that it could produce a huge amount of traffic. This is why many client interfaces generate a relevant amount of background traffic even if the user is completely idle, or they are not routing any searches[4].

Here we present a simple example based on version 0.6 of the Gnutella protocol specification. In Gnutella, the different messages are characterized by a given code in the `[PAYLOAD TYPE]` field; `PING` and `PONG` messages are characterized by values `0x00` and `0x01`, respectively. Figure 3.2 depicts the reference scenario, composed by six peers. Let us assume that the grey peer obtained a valid set of IP addresses from a proper bootstrap mechanism (in this case, if a Peer Cache similar to the one presented in Section 3.2.2. has been employed, $N_{Ep} = 2$).

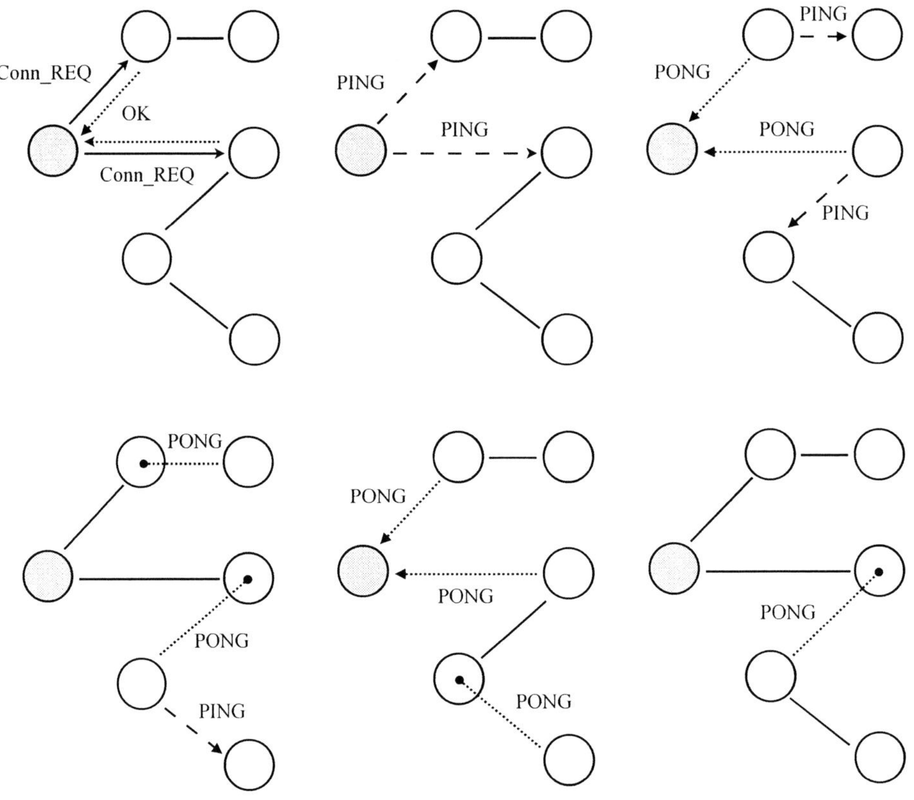

Figure 3.2. Evolution of a Gnutella-like peer entering the overlay (in grey) and discovering neighboring peers. Notice the PING-PONG mechanism and peers acting as information router.

As a first step, it tries to connect to the known peers by sending a *connection request* (depicted as Conn_REQ in Figure 3.2). Let us suppose that both nodes granted the request by sending a generic OK response. To discover other peers populating the overlay, as well as to keep track of changes due to churn, it probes its (unknown) neighborhood by

[4]We point out that a client interface may not route searches for the following reasons: i) searches are not taking place in a given timeframe; ii) searches are handled by a centralized component; iii) the peer acts as a leaf node in the overlay organization, i.e., it is not supposed to participate in routing queries and, due to transparency issues, the peer could not be employed to accept unsolicited incoming traffic, therefore not being able to cooperatively support some overlay operations.

sending PING packets. PING packets have also a TTL to limit the number of nodes to be contacted, for the same reasons already explained in Section 2.3.4., when we discussed how to perform searches in unstructured p2p overlays. Figure 3.2 portraits such evolution. Each peer receiving a valid PING message sends back a PONG to inform about its status. In addition, a peer is also responsible of routing back the PONG messages sent back from other peers. By repeating this procedure every given time step, it is possible to maintain the overlay consistent. To reduce the impact of churn, a node can leave the overlay by performing a gentle log-out operation, e.g., by informing its neighboring peers.

Notice how this continuos ping-ponging could generate a non-negligible traffic load, that in case of an inefficient overlay mapping, has to be moved for several times trough the same physical link (i.e., increasing the link stress as defined in Section 2.2.1.).

3.3.2. Server-assisted Methods

When a centralized component is deployed, it can be exploited to gather the status of the p2p portion and to issue proper information to peers to adjust the overlay. This is the case of BitTorrent, where the tracker is used to collect information about peers populating the distributed portion of the architecture. Besides, also the server-based portion of the eMule service relies on server assisted methods to update the lists of files, hence the peers active in the system.

This class of methods behaves similarly to the ping-pong one. Instead of producing a distributed flow, the central component sends PING messages to clients and updates their status accordingly (i.e., upon receiving a PONG before a proper timer expires). Again, properly exploiting log-out operations will reduce the possibility of mismatches among the reconstructed status (e.g., the one resulting after collecting PONGs) and the real one.

3.3.3. Structured Overlay Stabilization

The effect of churn accounts for inconsistencies in the finger table of peers composing the ring. In order to avoid a lack of efficiency or a high failure rate of look-ups, due to outdated fingers, the Chord algorithm implements a procedure called *stabilization*.

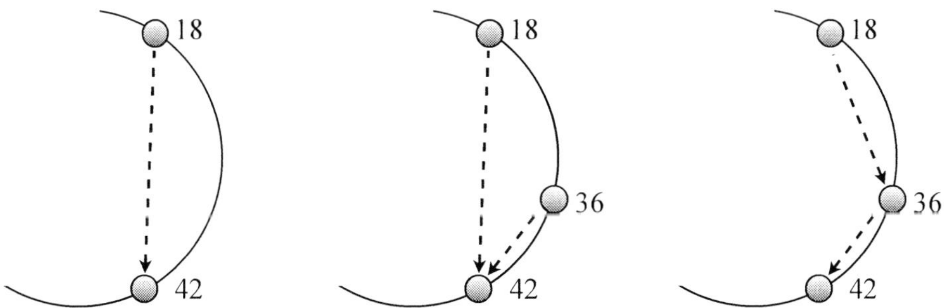

Figure 3.3. Stabilization of the Chord ring after peer 36 joins the ring. Notice that only one finger per peer is depicted, and key values are not present in the figure.

Figure 3.3 depicts a simplified[5] example of a stabilization procedure.

Let us suppose that peer 18 has a correct entry in the finger table pointing at peer 42. Then, a new peer (the number 36) joins the overlay, and updates its finger table for pointing to its successor, peer 42. In order to recover the inconsistence in the finger table of peer 18, it runs the stabilization algorithm, that basically forces the peer to find its successor again. By doing so, peer 18 updates its table and points, as a successor, to peer 36.

Stabilization is needed to assure that a look-up operation is successful, even in the middle of changes in the ring. In hugely populated overlays, changes during a look-up operation happen with a high frequency. However, if changes happen during a look-up, and before the stabilization procedure is complete, it is possible that a look-up will not fail.

There are three basic cases:

- all the finger tables exploited to perform a given look-up are valid (e.g., nodes with inconsistent finger tables are not contacted). Then, the look-up is successful and terminates, according to Chord, in $O(log(N))$ steps [21], where N is the number of peers composing the ring;
- all the successors are correct, but some finger tables are not updated: in this case, the look-up will be still correct, but slower;
- all the contacted peers have not the successors updated (thus, the stabilization has not been completed or performed yet), then the look-up may fail.

3.4. File Exchange Disciplines

File exchange disciplines are devoted to effectively move files or their parts from an host to another, typically by using TCP connections. Modern file-sharing applications implement sophisticated handshaking protocols and efficient transfer heuristics to achieve benefits due to the presence of multiple copies of files in different remote peers. In this perspective, several helping techniques are needed. For instance, methodologies to precisely identify a file (or its subparts), protocols to exchange information about files' locations and mechanisms to policy the multiple concurrent transfer requests.

3.4.1. Chunking and Hashing Mechanisms

First incarnations of file-sharing applications identified files via their names, and simple descriptive parameters, such as the quality adopted in the encoding process (i.e., the bit-rate). Such a scheme is simple but it introduced some hazards: for instance, an user can maliciously rename a file to cheat other users.

The major flaw consists in the impossibility of establishing for sure if two files are equal. This fact impeded to exploit efficiently *multiple download* disciplines, i.e., retrieving the file concurrently from different users. In addition, to rationalize this process, a file is subdivided into different parts, commonly called *file-chunks* or simply *chunks*. A discussion about the

[5]Specifically, key values are not mentioned here, as well as key movements across churning peers. Interested readers could find more details in [21].

performance gains achievable with such a method will be presented in Section 7.6.. If needed, chunks could also be further subdivided in smaller sub-parts to allow a more "fine grained" file management. In order to identify files and their sub-parts, the concept of *hash* has been introduced.

Subdividing a file into smaller parts enables to perform the operation called *swarming*, which is the ability of downloading a file concurrently from different remote peers. Then, like a "swarm", different file chunks are retrieved from remote peers. Besides, the search mechanism available in the overlay also allows to exploit the *hording* technique, which consists in the ability of exchanging information to retrieve chunks of a file.

Regarding the *hash*, it is a kind of digital signature that is unique for a given file or, more generally, for a given block of data. Typical disciplines rely on applying an hash function to the stream of byte composing the file, in order to have a unique identifier. Commonly employed hash functions are those based on the MD4 [30] algorithm[6]. This allows to identify precisely a given content, despite the associated name in the file-system. Different file-sharing services have different hashing mechanisms. In Chapter 6 we will discuss the mechanism adopted in eMule, while in Chapter 7 we will analyze the one based on meta-data adopted by BitTorrent.

3.4.2. File Handling

Despite the core functionalities devoted to exploit the overlay, a major component is the one related to the handling of files.

According to the peculiarity of a given file-sharing application, the client interface may/may not have features to manage the files to be shared. For instance, this is the case of the original BitTorrent implementation, where the client interface only handles one file at time. Such features are mostly related to how a user's file collection is organized on the local file-system.

More critical features are related to how the files, intended as "public resources" are managed by the overall distributed architecture. The most important point break between Napster and its successors was the introduction of chunk-related mechanisms. In fact, Napster allowed to download a file sequentially, from the beginning to the end. In addition, it was also possible to resume the download if interrupted.

Alas, this mechanism had some drawbacks, specifically:

- mismatch errors: due to the "continuous stream" nature of a transferred file, it was common that resuming a download produced some errors. A mismatch is an inconsistency between the downloaded data and the original one. Thus, it is not possible to properly "attach" the received stream into the already received amount. This hazard has been recovered with a technique called *trimming*[7], but it was effective only for a small subset of mismatch errors;

---

[6]If you have access to an Unix machine, try in the shell `md5` *filename* to obtain the hash of a file, even if calculated with the MD5 algorithm, it gives idea about what an hash of a real file looks like.

For instance, the MD5 of the .pdf draft of this book, computed by issuing the command `md5 Book.pdf` in a Unix terminal, is `6d7710c609f914b9ff09936045bf33ab`.

[7]The trimming technique consists in counting the bytes obtained so far and if odd, the exceeding ones are dropped in order to have an even multiple amount of bytes.

- error recovery: handling the file in a monolithic way impeded to recognize errors until the last bit composing the stream had been received;
- "short-circuiting": since a file was available to the system only upon the complete reception, it was not possible to deliver available bytes of the incomplete file while receiving the missing ones;
- incomplete availability: as said, every peer sent and received the file sequentially, thus it was common to have many incomplete copies of the file to be exchanged within the system.

To cope with such pitfalls, developers introduced mechanisms to abandon the monolithic nature of a file in favor of handling it as a superposition of smaller parts. This mechanism offered some major benefits, while it added complexity in the client interface, and thus the need of some file handling mechanisms and an enriched logic devoted to compute the hashes. For instance, each file must be processed in order to: i) calculate a map describing the chunk subdivision and ii) computing the hash for each chunk and the overall file.

Besides, the major benefits are:

- enhanced level of transfer parallelism: it is possible to send and retrieve different chunks concurrently from different remote peers. This allows to better exploit the available bandwidth;
- fine-grained scheduling and policies: typically, in file-sharing applications each peer requests one file. Since the upload bandwidth is not infinite, it is only possible to serve a finite amount of remote peers simultaneously (as it will explained in Section 3.4.3.). Thus, policing the queue of waiting peers by using the chunk granularity instead of the one at the file level, it is possible to better exploit available resources and building also some "high level" scheduling policies;
- avoiding long-time locks: if files are treated "atomically", it is possible to act in a queue configuration where peers are downloading very large files (e.g., with a size of several Gigabytes), thus preventing other peers to be served for a long time;
- allowing "short-circuiting": since a file is subdivided into independent portions, it is possible to deliver available chunks before the completion of the overall original file;
- avoiding error propagation: each chunk is received and then verified. If errors occur, the chunk (or a sub-part) is not propagated anymore. Thus, it is possible to re-ask only for a given chunk and circumscribing errors;
- increasing the file availability: instead of having many replicas of a incomplete file stuck at the same level of completion, by distributing in a "smart" way the chunks composing the original file, it is possible to have a distributed copy. In this vein, even if not completely available at one peer, it is possible to retrieve different chunks in different peers. Heuristics aiming at this result will be investigated in Chapter 5.

3.4.3. Queueing

In order to manage the upload bandwidth, p2p file-sharing applications are used to subdivide the available resource in K upload slots. Each slot is then utilized to transfer data to a remote peer. However, to keep the implementation feasible and to avoid bandwidth trashing due to overheads, K must be finite. Then, each peer serves up to K remote peers, and exceeding ones are queued. For this reason, it is important to implement some queuing management mechanisms.

Past systems, such as WinMX, had the queue managed in a manual fashion, thus giving to users the freedom of overruling the natural queue's evolution. This approach has been dropped and modern client interfaces exploit an automatic queue management. It is interesting to note that, having peers remotely queued allows to exploit disciplines to reward/punish some peers' behaviors. This is why, queue management is often tightly coupled with the enforcing heuristics portion of the file-sharing architecture, described in Section 3.5.2.. As a successful example, Chapter 6 will deeply discuss the eMule queuing architecture and its exploitation for enforcing cooperation.

The upload queue could also be populated by a huge amount of remote peers, for several reasons. Specifically: i) a peer shares a huge amount of files, or an highly interesting file selection, thus becoming contacted by a relevant amount of users; ii) the available upload bandwidth could not suffice for promptly satisfying the volume of requests, thus exceeding peers are queued. Consequently, peers could remain in a remote peer's queue for a long time. Peers are free to leave the system at a given time in an unpredictable fashion, thus each peer must update their queue accordingly and continuously. This accounts for the need of developing an adequate threading architecture within the client-interface and reflects in signaling traffic over the network to recognize if queued peers are still in the system: that is another effect of the churn.

Depending on the degree of sophistication of the client-interface, K could be fixed or adjusted according to some heuristic, thus becoming the number of upload slots time varying. In this case, we define as $K(t)$, the number of slots at a give time t. As it will be discussed in Chapter 8, there are also cases when multiple queues are in place, thus having a K_i, $i = 1, \ldots, Q$, where Q is the number of available queues.

3.5. Other Functionalities

In this section we analyze other functionalities needed to guarantee the proper operation of a file-sharing system. Such features are as important as those previously explained; however, we will discuss them briefly since they will be investigated deeply in the rest of the book.

3.5.1. Search Handling

Peers participating into a file-sharing system are *often* responsible of performing and managing searches for a given content. The adjective *often* has been introduced since, in some applications, searches are delegated to a third-party infrastructure. This is the case of BitTorrent, where searching for a given resource is shifted to searching for a specific metadata file, called the `.torrent` file. Usually, this action is performed via the World Wide Web (WWW), or by using special purpose repositories.

Anyway, the client interface must be able to generate and receive queries and results. In addition, if a centralized component is not in place, each peer must be able to cooperatively participate in evaluating/routing the query. This is the case of Gnutella, as explained also in the example presented in Section 2.3.4..

3.5.2. Enforcing Heuristics

Since p2p file-sharing systems heavily rely on the active participation of peers, some enforcing heuristics are needed. Basically, two major resource contributions must be taken into account: i) enforcing a fair use of the available bandwidth and ii) promoting the active sharing of files.

Actually, such problems are solved by implementing simple policies within the client interfaces during the set-up phase, thus by means of a set of distributed heuristics to avoid "disruptive" behaviors from peers. Many systems also employed centralized heuristics to force cooperation: such topics will be discussed in detail in Chapter 5, while optimized procedures have been also investigated in the literature, and they will be presented later in Chapter 8. In addition, Chapter 9 discusses the effectiveness of such heuristics by investigating data collected from a real environment.

3.5.3. Assisting Communication among Peers

Despite the logical decoupling introduced by the overlay, the presence of the underlying network infrastructure accounts for several hazards. The main problem is due to the reduced end-to-end transparency introduced by devices such as firewalls and Network Address Translation (NAT) devices. In order to recover to this particular settings in a transparent manner, about the totality of file-sharing applications provide techniques to assist communication among peers. In Chapter 4, we discuss both solutions exploiting techniques of NAT traversal and cooperation through peers.

3.5.4. Companion Services

Companion services are not strictly related to file-sharing functionalities but are often employed to enrich the user experience. We cite, among the others, chat services, instant messaging and files preview. Following, we briefly discuss the integration of basic web services within standard client interfaces.

Sometimes the files shared are not the ones you will expect. Many users do alter the name of a file for impressing other ones. Such files are called "fakes". Even if, when in presence of hashing mechanisms (see Section 3.4.1.) the file is uniquely identified despite its name on the local filesystem, this fact can force users to misbehave. Thus, to avoid downloading a fake, some companion services implemented within client interfaces are proposed.

One of the most popular is to build a "repository" of the file history (e.g., all the different names of a file during its lifetime) and update it on the fly. If a user recognizes a fake, he/she can signal it and the fake database will be updated after a check. Many programs, such as eMule, can be integrated with the web via a kind of webservices. For instance, it is possible to instruct the eMule client interface to verify the identity of a file by checking the web as follows:

```
http://www.example.com/search?p=ed2k:#filesize:#hashid
```

where `http://www.example.com/` is the URL of the search engine for retrieving information about filename history, `filesize` is the size of the file under investigation and `hashid` is the hash associated to a particular file.

As it will be explained later, in Chapter 7, metadata-based systems are quite immune from this kind of misbehavior, thus relieving the client interface from this kind of features.

Chapter 4

Dealing with the Network

The increasing complexity of the Internet accounts for a more challenging network scenario where file-sharing applications have to operate. Summarizing, the main issues are: time varying IP addresses assigned to end-nodes (e.g., by employing DHCP-based disciplines) and the lack of end-to-end transparency due to mediating devices such as NATs and firewalls. Concerning dynamic addressing, it mainly impacts over the selection of the first node to contact to join the overlay. In this perspective, we discussed such issue in Section 3.2..

Therefore, this chapter deals with countermeasures and techniques that have been adopted to cope with the specific problems arising from the lack of transparency accounted by the presence of mediating devices. Specifically, we address the usage of basic NAT traversal techniques and of overlay-based relaying solutions also for providing a simple and quick method to avoid issues arising by the adoption of firewalls. Lastly, a quick showcase of the recent mechanisms for automatic configuration of mediating entities is provided.

4.1. Why Traversal Techniques are Needed

The Internet has evolved since its birth, and the original end-to-end vision has been slowly mutated due to different reasons. We mention, among the others: i) the need of recovering to the depletion of IPv4 addresses, for instance by "multiplexing" a small subset of public IPv4 addresses in favor of a larger customer base equipped with private IPv4 addresses; ii) the quest for increasing security spawned the adoption of NATs to secure networks areas as well as the diffusion of firewall devices; iii) the envisaged progressive adoption of IPv6, which will restore the end-to-end transparency, will be smooth and lengthy, then IPv4 to IPv6 (and viceversa) translation devices will be in place for an undefined time frame.

Solutions such as the Classless Inter-Domain Routing (CIDR) [31] appears to be only valid for the short-term period. In addition, a developer should assume the presence of such devices for an undefined time-frame. As a consequence, facing the lack of transparency is mandatory to engineer successful and long-living file sharing applications.

4.1.1. Impacts on the Overlay

As hinted before, the key issue introduced by the aformentioned devices concerns the reduction of the end-to-end transparency, that is a critical propriety of the underlying network deployment, especially for providing connectivity to highly cooperating entities located on the border of the network. Without such a desirable property, the resulting overlay could be endangered, since it is mainly operated by end nodes, as well as the effective exploitation of the cooperative efforts at the basis of p2p file-sharing applications. For instance, a peer willing to actively contribute resources to the file-sharing network, could not be able to participate due to the lack of transparency. This may cause some misbehaving of such mechanisms devoted to foster and quantify cooperation among peers, as it will be explained in Chapter 5 and further proved by the analyses presented in Chapter 9.

Then, in order to engineer effective applications is important to implement some countermeasures to deal with the presence of NAT devices.

4.2. NAT Traversal

Many traversal techniques have been developed and some of them have been also stuffed in the IETF standardization pipeline. The most simple but effective one is called *hole punching*, that has been primarily introduced for pushing UDP traffic over a NAT.

Recent advancements allow to use the hole punching mechanism also to provide a traversal solutions for TCP traffic. Due to core differences between the two protocols, pushing TCP traffic through a NAT is more complex than pushing UDP. Moreover, the precise behavior of "punched" TCP traffic, and how different NAT devices processing it, are still not completely understood.

In fact, NAT implementations are quite heterogeneous, in terms of functionalities and basic behaviors in processing the traffic. This reflects in the major issue accounting for the reduction of the effectiveness of the presented approaches: *not all the NAT devices behave in the same manner.*

In other words, a developer could encounter a well-behaving NAT as well as a broken one. In addition, a complex cascade of NAT devices could be present hiding/joining different portions of a Wide Area Network (WAN) or geographically distributed Local Area Networks (LANs). Lastly, also the underlying OS hosting the p2p file-sharing application plays a role. It is important that the available protocol stack (i.e., the TCP/IP implementation of the OS) supports a well behaving TCP implementation (e.g., its protocol stack is able to provide multiple connection opens) and allows programmers to tweak some protocol behavior (e.g., adjusting the value of the IPv4 TTL on-the-fly)[1].

The rest of this discussion is based on two different sources: seminal papers and IETF standard documents. To allow readers to deeply investigate the topic in an autonomous way, NAT basics will adhere to IETF standard documents. Besides, the proposed traversal techniques are explained by borrowing some concepts from the seminal works of Ford, Srisuresh and Kegel [32], and Guha and Francis [33]. Readers interested in the field of development are encouraged to read such works, since the Authors have also released the

[1] If IPv6 is employed, the needed field to be adjusted is the *Hop Limit* one.

code utilized to experiment with NAT traversal punching techniques. Moreover, the proposed sources are also valuable to understand the status of commercially available NATs, since results have been mostly collected from devices deployed in "the wild".

4.2.1. NAT Basics

NAT devices have been under the attention of many network device vendors, and spawned a thorough discussion within the standardization community. The IETF has produced many documents and the interested reader should at least read [34] to gain a minimal comprehension of NAT-related internals and the basic terminology. In addition, [35] should be read to gather a general vision of what the duties and capabilities of a NAT box are.

Roughly, a NAT is a device able to translate the addresses within a IP datagram to other ones, in order to guarantee the correct data delivery between private and public address realms and viceversa. This feature is commonly exploited to map a wide population of private IP addresses into a smaller one of public IP addresses.

For instance, it is common to have some NAT software allowing to do connection sharing in home networks. Thus, the unique public IP addresses given by the ISP is shared across multiple devices to simultaneously employ the Internet connection from the different entities populating the network.

Even if this is a very reductive description of NAT capabilities, and internals acting in the "background", it is one of the most widespread, thus reflecting the so called Small Office Home Office (SOHO) network environment. Nevertheless, NATs are increasing their diffusion over ISPs, for instance for assuring Internet connectivity to a wide population of users arranged in a nation-wide private realm[2].

As mentioned, the overlay employed at the basis of p2p file-sharing applications offers isolation properties, but there is a place where a "contact point" between the internetwork of virtual links (i.e., TCP connections) and the underlying network architecture exists. Put briefly, the overlay relies on the possibility of establishing point-to-point TCP connections; besides, UDP traffic could be employed to convey extra signaling, for instance exchanging in a cooperative way information about available resources (see for instance the hording feature, discussed in Section 3.4.1.). In this vein, also for p2p applications the concept of *session* is important as well as in the NAT world.

Before discussing traversal techniques, let us consider two definitions introduced within the IETF standard documents.

Session endpoint: a session endpoint for TCP or UDP communications, is the 2-tuple:

`(IPAddress,PortNumber)`

[2]As said, NATs are used also to secure users belonging to an ISP. But, a recent trend is to use NAT to impede users to aggressively use the network. For instance, with the increasing availability of full-duplex fiber accesses at low fees, residential users could act as a "small" ISPs, i.e., hosting services and contents for third parties. Even if such scenarios could be avoided by using "lawsuit-countermeasures", the history of p2p file-sharing has extensively proved that this could not suffice. Then, offering a very high-speed Internet access through a NAT highly reduces the possibility of a user "sub-selling" its bandwidth. In addition, the usage of NAT is also exploited as a palliative to the aggressive usage of file-sharing applications, sometimes jointly with traffic filtering. Notice that traffic filtering is somewhat perceived by users as a "display of power", while relying on NATs is felt as less invasive and mainly related to how the network has been architected.

Session: a session is identified by the 4-tuple:

`(LocalIPAddress,LocalPortNumber,RemoteIPAddress,RemotePortNumber)`

Notice that a session is basically the composition of two session endpoints.

To be able to completely describe the produced traffic, also the concept of *direction* is needed. The IETF defines the direction of an UDP flow as the direction of the first transmitted UDP datagram.

In essence, NAT devices act on the aforementioned tuples to allow users without a public IP address to connect to public IP realms, by translating private endpoints into public ones. The way this translation is performed characterizes different flavors of NAT devices.

The detailed discussion about different flavors of NAT devices available is out of the scope of this book, but we cite, among the others, two main categories of devices: basic NAT and Network Address - Port Translation (NAPT) one. The former translates only the L3 addresses, while the latter translates also port numbers, i.e., the entire session endpoint.

4.2.2. UDP Hole Punching

The technique called *UDP hole punching* allows two peers to set up a direct UDP session with the help of a well-known third entity, commonly an ad-hoc server. This mechanism also works when both peers lay behind different NATs. Let us assume that both peers are connected with a *rendezvous server* called S. As an additional hypothesis, we assume S able to retrieve the session endpoint belonging to the requesting peer. For instance, this could be achieved by using a proper signaling mechanism, or by embodying such information within the data exchange performed for registering with S.

Before proceeding, let us introduce some formal definitions. Let us assume that two peers, P_A and P_B, laying behind a NAT, want to communicate. Moreover, let us define two different kinds of endpoints. Define E_{P_A} as the endpoint that P_A believes to use for communicating with S: obviously this is a private endpoint, such as $E_{P_A} = (IP_{P_A}, Port_{P_A})$, where IP_{P_A} and $Port_{P_A}$ are the private IP address and the local port number of peer P_A, respectively.

Also define as $E^S_{P_A}$ the endpoint that S actually observes: owing to the presence of the NAT, this is a public endpoint. Define $N()$ as a translation, i.e., a function:

$$E_x = N(E_y)$$

that is the mapping of the endpoint E_y into the endpoint E_x. Notice that the NAT properly works if it is possible to do the viceversa, i.e., N^{-1} exists. Thus, each encountered NAT performs an address translation and, in the case when only one NAT is encountered, the resulting translation is: $N(E_{P_A}) = E^S_{P_A}$. Then, $E^S_{P_A}$ has the form $E^S_{P_A} = (IP^S, Port^S_{P_A})$, where IP^S and $Port^S_{P_A}$ are the public IP address of the NAT and the NAT-assigned port number observed by S after the translation performed by the NAT, respectively.

Obviously, the same applies for P_B.

With such preliminary knowledge, let us describe the UDP hole punching technique. Basically, it is exploited via the following three steps (as presented in [32]):

1. P_A does not know how to reach P_B. Then, P_A requires assistance in establishing a session to S;

2. S sends to P_A both the known endpoints for P_B, i.e., E_{P_B} and $E^S_{P_B}$. Meanwhile, S sends to P_B the connection request issued by P_A and sends both the endpoints of P_A, i.e., E_{P_A} and $E^S_{P_A}$. At the end of this process, both P_A and P_B are aware of all the available endpoints;

3. As soon as P_A receives the P_B's endpoints it starts to send UDP packets to both endpoints simultaneously. Meanwhile, it "locks" with the P_B's endpoints, from which receives a valid response first. The same does P_B.

Now, let us suppose that both P_A and P_B lay behind the same NAT. In this perspective, they are sharing the same private IP address realm. Both clients have established a session with S. Then, let us apply the aforementioned UDP hole punching technique, allowing P_A to start an UDP communication (session) with P_B. P_A sends a connection assistance request to S, thus it becomes aware both of E_{P_B} and $E^S_{P_B}$. Then, S forwards to P_B both E_{P_A} and $E^S_{P_A}$.

Both peers start to send datagrams each other, i.e., the following UDP flows to all the known endpoints are generated:

1. $P_A \rightarrow E_{P_B}$, generating the following session: $(IP_{P_A}, Port_{P_A}, IP_{P_B}, Port_{P_B})$;

2. $P_A \rightarrow E^S_{P_B}$, generating the following session: $(IP_{P_A}, Port_{P_A}, IP^S, Port^S_{P_B})$;

3. $P_B \rightarrow E_{P_A}$, generating the following session: $(IP_{P_B}, Port_{P_B}, IP_{P_A}, Port_{P_A})$;

4. $P_B \rightarrow E^S_{P_A}$, generating the following session: $(IP_{P_B}, Port_{P_B}, IP^S, Port^S_{P_A})$.

As regards UDP datagrams generated by sessions 2 and 4, they may or may not reach the destination, depending whether the NAT supports the hairpin translation[3] [32]. Conversely, datagrams generated by sessions 1 and 3 reach their destination. In addition, messages routed directly through the private network are supposed to be delivered faster than those routed through the NAT device. As a consequence, P_A and P_B will utilize private endpoints to start next regular communications.

We point out that the UDP hole punching technique also works when in presence of peers behaving through different NATs or when in presence of multiple levels of NAT. Such scenarios are out of the scope of this book, but interested readers can find a detailed discussion in the seminal work available in [32] and in the references therein.

4.2.3. TCP Hole Punching

As said, the TCP hole punching technique is more complicated and less understood than the UDP counterpart. The concepts related to session endpoint and session are still the same as the case of UDP data. However, in the case of a TCP session, the direction of

[3]The *hairpin translation* (often called also *loopback translation*) allows an host on the private network behind the NAT to communicate with other hosts on the same private network using public (translated) port bindings assigned by the NAT.

the data flow is given by the direction of the first `SYN` packet. The main drawback of this technique is that it is less supported by NAT devices than the UDP hole punching.

As discussed in [32], the key challenge of TCP hole punching does not reside in the specific protocol, but in the implementation of the Application Programming Interface (API) that supports TCP communications, i.e., BSD-Sockets[4]. In fact, in order to be effective, hole punching must be corroborated by the possibility of using a single local TCP port for listening to incoming TCP connection requests and to initiate multiple outgoing TCP connections simultaneously. This is consistent with the behavior of the UDP hole punching, where UDP streams are sent to both endpoints concurrently.

Currently, many operating systems support the `SO_REUSEADDR` option allowing to use multiple `socket( )` system calls on the same endpoint. However, this is not sufficient, since also the `SO_REUSEPORT` option must be available.

TCP hole punching is then very similar to the UDP variant and interested readers will find more detail in [32]. We only report here that it exists also a particular flavor called *sequential hole punching*. In sequential hole punching, for instance implemented in the NaTrav [36] software library, peers connect one to each other sequentially rather than simultaneously. Such procedure is slower than the one based on parallel TCP connections, but it allows peers to exploit hole punching even if not able to set the `SO_REUSEADDR` option. Readers interested in gaining more comprehension of TCP/IP network programing are encouraged to read the classical textbook by Stevens [37].

4.2.4. A Remark on NAT Devices

As explained, NAT traversal techniques are very effective but they heavily rely on the specific implementation of the NAT. In fact, even if correctly exploited, many devices do have misbehaving software implementations preventing the proposed approach to actually work. In [33], Authors report that many devices are used to mangle data packets, e.g., by adding an offset to the original sequence number of TCP segments. Besides, also the correct implementation of the complete TCP state machine within the translating device is crucial to successfully exploit the proposed mechanism.

In a nutshell, the TCP traversal technique appears more coupled on the specific NAT implementation rather than the UDP one. This is one of the reasons why many p2p application developers do prefer to rely also on UDP for file transfers, that are operations typically exploited by TCP and harder to manage when UDP is used.

Then, a developer who wants to engineer a p2p file-sharing application must face a complex set of problems. Nevertheless, developers must become aware that NAT devices exist in different flavors, and this additionally increases the required effort to provide a client interface able to operate in very different contexts.

Summarizing, the different categories of NAT devices are:

- *Full Cone NATs*: all requests originated from the same private session endpoint are translated into the same public session endpoint;

[4]If you have access to a Unix machine, you can retrieve more information about the BSD-Sockets API for the ANSI-C language by accessing the man pages, e.g., by typing `man 2 socket`.

- *Restricted Cone NATs*: they have the same behavior of full cone NATs when translating packets, but they do not forward unsolicited data packets received from the public realm. This augments the level of privacy of the private portion of the network, but also increases the effort needed to establish p2p communications;
- *Port Restricted Cone NATs*: they impose restrictions not only on the IP addresses, but also on ports used to translate session endpoints;
- *Symmetric Cone NATs*: they introduce more restrictions than the Port Restricted Cone NATs, since they perform a new mapping for each different connection, thus introducing tight restrictions in the process.

Summing up, the exploitation of NAT traversal techniques strictly depends on the nature of the device itself.

4.3. Connection Reversal

The *connection reversal* is a simple and well known technique used to establish a communication between two peers when a NAT is deployed somewhere in the middle of the network. Nevertheless, it is also extensively used when, instead of NAT devices, some firewalls impede the required traffic flow.

Such technique has been successfully employed in the Kazaa overlay [38], as well as in other file-sharing systems. Figure 4.1 depicts the connection reversal procedure in a fictitious unstructured overlay with superpeers (as in the case of Kazaa). Relying on superpeers (supernodes) accounts for another benefit of the adoption of an overlay composed by two class of peers, i.e., a 2-level overlay organization. In fact, supernodes are supposed to have a "transparent" Internet access, thus being able to seamlessly interact with other peers. Nevertheless, many p2p file-sharing applications, such as WinMX, put also constraints on hosts' bandwidth requirements to act as a supernode, thus assuring that a peer responsible of handling the connection reversal is never overloaded, or has its bandwidth swamped.

Notice that this approach is also well suited when in presence of hybrid overlays; in fact, the centralized component could be exploited to handle the needed signaling devoted to start the connection reversal.

We point out that the connection reversal technique is not solely employed for p2p applications, but it is a general framework to cope with NATs without resorting to the hole punching techniques. As reported in [32], the connection reversal technique is then often exploited by deploying a server to handle the needed signaling. Let us analyze in detail the procedure at the basis of the reversal technique. We refer to the reference scenario depicted in Figure 4.1. As an hypothesis, let us suppose that both peers are connecting to the same supernode, which also has the needed logic to handle connection reversal requests. If the peer behind the NAT wants to start a communication with the remote peer, then no problems arise. However, due to the presence of the NAT device, the vice-versa is impeded. Thus, the peer sends a *reversal request* to the superpeer, which has an already established TCP connection[5] with the other peer. Then, the superpeer exploits such connection to relay

[5] We recall that TCP connections are bidirectional.

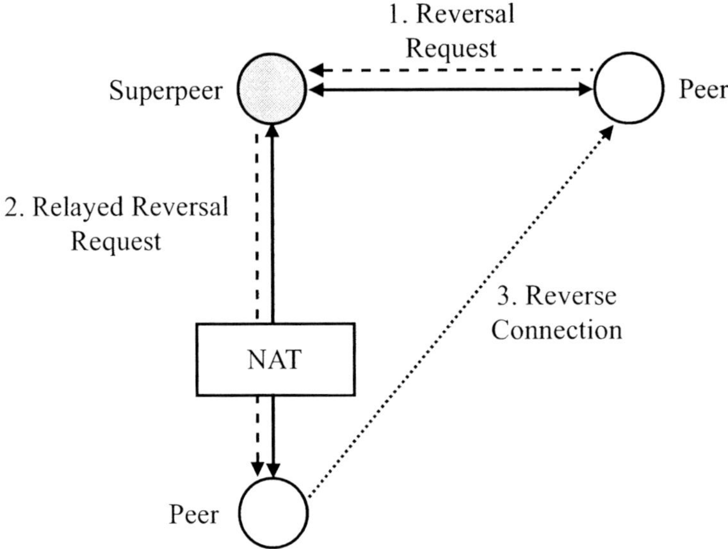

Figure 4.1. The *connection reversal* procedure applied by cooperating peers in the overlay to traverse a NAT device. Solid lines are already established TCP connections, while dashed lines are the signaling flows employed to request a connection reversal.

the reversal request. The target peer receives the reversal request and starts a direct p2p connection with the requesting peer.

4.4. Universal Plug and Play

The Universal Plug and Play (UPnP) [39] is an automatic mechanism developed by Microsoft (`http://www.microsoft.com`) to support the configuration of network devices without the need of user intervention. The UPnP is built on top standard IETF protocols. However, it mainly exploits the HTTP over UDP, which is not a standard mechanism, rather a particular composition of two already available standards. UPnP is also capable of configuring a well-recognized category of network devices. For instance, it is employed to set-up Internet gateways, via the Internet Gateway Device (IGD) Standardized Device Control Protocol [40]. A detailed discussion about UPnP is out of the scope of the book, since it is evolving very vast.

Anyway, we cite it in this chapter for the following reasons:

- as a consequence of the increasing complexity of network devices (for instance in terms of configurations, the need of mapping ports, ...) configuring some aspects of a file-sharing application is becoming complicated for non skilled users. But, p2p file-sharing applications take power from “the masses”, thus relying on automatic configuration mechanisms allows to reach a huge user population;
- many client interfaces of the most popular file-sharing applications (i.e., eMule and

BitTorrent) already support the UPnP to properly configure firewalls and NATs to avoid transparency problems;

- UPnP offers mechanisms for NAT traversal and could be another solution besides the traversal techniques previously discussed in this chapter, and other ones, such as the Traversal Using Relay NAT (TURN) [41] or the Simple Traversal of UDP through NAT (STUN) [42];
- UPnP could bring to security risks when interacting with network devices. In this perspective, its adoption in p2p file-sharing applications, which generate aggressive traffic patterns and where thousands of nodes get contacted, thus "spreading" reachability information through the Internet, should be carefully evaluated.

Chapter 5

Organizing the User Anarchy

Alas, the adoption of the p2p as an enabling technology for the development of sophisticated file-sharing applications does not come for free.

An important issue concerns the total freedom of users in assigning resources to client interfaces and their attitude of sharing contents. Nevertheless, users may leave the file-sharing network at any time, for instance disconnecting upon completion of a needed content (i.e., files to be downloaded), even if a proper degree of reward to the rest of the peer population has not been hit. In addition, peers leaving the system in an unpredictable manner can reduce the efficiency of the overlay. Summing up, the anarchic nature of users could impede the system to efficiently exploit available resources.

This chapter deals with ideas and solutions adopted in modern file-sharing applications to cope with the aforementioned issues. Instead of proposing a discussion at large, we will focus only on general concepts available in modern client interfaces; then, we will specialize such concepts in Chapter 6 and Chapter 7, when presenting eMule and BitTorrent, respectively. Specifically, we deal with the credit system and queue modifiers, the bandwidth enforcers, distributed heuristics and techniques based on peers' age.

5.1. On the Need of Restoring Order

In p2p file-sharing systems, like other p2p-based environments, peers are responsible of cooperating for maintaing the overall service infrastructure up-and-running. Besides, file-sharing services heavily rely on files and bandwidth availability. Then, it is important that users configure their client interface with enough bandwidth to support the service; nevertheless, they must share some contents. In addition, the churn affecting the system could impede the exploitation of resources or accounts for temporary file unavailability.

The sum of egoistic behaviors, emphasized by the huge population of peers, could endanger the overall file-sharing system. The freedom of users must be balanced by their responsibility. But, responsibility of end users is something not obvious nor sure. Therefore, virtuousness must sometimes be incentivized, if not forced or imposed.

For such reasons several mechanisms have been developed to force users (thus, peers) to cooperate, and to balance the intrinsic anarchy of systems where users have quite a total freedom.

5.2. Credits and Queue Modifiers

Credit system and queue modifiers are usually employed in synergy, thus resulting in being tightly coupled mechanisms. They have been firstly introduced in file-sharing applications to cope with three different drawbacks, namely:

- *avoiding selfishness*: in order to prevent selfish actions, credit systems have been introduced as a possible countermeasure to enhance the robustness of the p2p file-sharing application and to increase both file availability and the overall service capacity;
- *lack of persistence of users' behavior*: an user can strongly change his/her behavior from session to session. However, relying on credits allows to easily keep track of the early history, to reward past efforts or to prevent the reiteration of non-cooperative actions;
- *avoiding manual queue management*: legacy services forced users to manage queues manually. On one hand, the task was boring, while, on the other hand, it allowed users to overrule decisions[1] taken by the client interface, hence invalidating any enforcements. In addition, the users' actions are often performed on a per friendship behavior, rather than taking the decisions on real degrees of contribution.

5.2.1. Credits

Credits are used to quantify in an objective manner the degree of contribution of a peer with respect to the rest of the population of the file-sharing service. In particular, for this application, credits are often related to the provided data volume to another peer, e.g., they quantify how a peer contributes to the overall exchanged volumes. Having a precise performance index to describe the behavior of a peer allows to exploit reward/punishment operations, in order to force peers to act nicely with respect of the entire system. Even if this does not completely prevent anarchy, (e.g., users are still able to manually set some parameters or to share only few files), it is possible to use credits as a way to feed algorithms that influence the behaviors of other peers. As an example, if a queuing system is adopted, it is possible to use the credits to shift the position of a given peer within another peer's remote queue.

Let us define, in a general way, what a credit is.

A credit achieved by a peer, at a time t of calculation, is a function $C(t)$, that quantifies the behavior of a peer[2]. For instance, a generic credit score could be the amount of uploaded data to a remote peer up the time t.

Notice that the lack of transparency of NAT devices (as discussed in Chapter 4) or mediating entities (e.g., Performance Enhancing Proxies employed to improve performances of TCP transfers in satellite-based accesses [43]) could impede a peer to properly cooperate, thus reducing the amount of exchanged data to be employed to earn/spend credits. This is

[1] Even if legacy client interfaces were not provided of queue management disciplines, a "natural" one was the straight First Come First Served (FCFS) policy.

[2] Notice that a proper $C(t)$ has to be evaluated for every peer in the system, i.e., $C(t) = C_j(t)$ for $j = 1, \dots N$, where N is the number of peer in the system. However, to avoid burdening the notation, we drop the dependence from j.

why traversal solutions are important not only for establishing connections for creating the overlay, but also to not endanger the fairness and the effectiveness of the overall file-sharing infrastructure.

5.2.2. Queue Modifiers

As said, file-sharing applications use the available upload bandwidth to serve remote peers. Obviously, bandwidth is not infinite and to augment the overall system efficiency, peers are served concurrently; thus, the available bandwidth is subdivided into a finite number of upload slots. Let us define S as the number of available upload slots. To avoid a reduction of the perceived quality, and to mitigate the overheads, S must be carefully selected. In some cases, S could also be dynamically adjusted, for instance to adapt the service rate to bottlenecks between the sender and the receiver, thus reflecting in a function of time $S(t)$.

Anyway, the finite nature of S reflects in the need of queuing exceeding peers: in other words, peers having a position in the queue greater than $S+1$ are queued. Then, managing the first S position in the queue, is a crucial operation that can bring the entire file-sharing service into an inefficient behavior, or at least influence its fairness.

In order to guarantee a fair and automated management of queued peers, a popular approach is to use the information provided by the credit score to feed a specific function, called *queue modifier*, which allows to shift the position of a peer in the waiting queue according to its behavior.

Let us define a queue modifier at the time t of the calculation, starting from the "first contact", as a function

$$m(t) = g(\tau(t), \mathcal{C}(t))$$

where $\tau(t)$ is the time spent in queue at time t by a peer waiting for being served and $\mathcal{C}(t)$ is the credit as defined in Section 5.2.1. to be used to achieve a better position in the queue[3].

Then, define

$$\tau^*(t) = m(t) * \tau(t)$$

as a fictitious time used to compute the actual rank in the queue, hence allowing to either promote or penalize queue advancements, according to each peer's behavior.

Figure 5.1 depicts a toy example of a queue managed via queue modifiers. The example shows peers queued according to their "age" (i.e., by sorting their arrival time τ_i), and then further queued by the fictitious time computed taking into account modifiers (i.e., by sorting themselves using the τ_i^*).

Notice that the function $m(t)$ varies according to the specific file-sharing application; in Chapter 6 the modifier strategy adopted by the eMule system will be deeply investigated, hence we "specialize" the $m(t)$ for a real-world system.

[3]Again, notice that a proper $m(t)$ has to be evaluated for every peer in the queue, i.e., $m(t) = m_i(t)$ for $i = 1, \ldots, N_q$, where N_q is the number of peer populating a queue. In this perspective, the same applies to τ_i and τ_i^*. However, to avoid burdening the notation, we drop again the dependence from i.

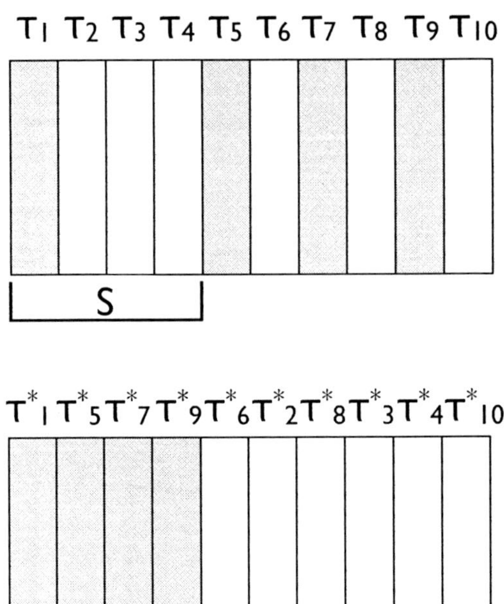

Figure 5.1. An upload queue with $S = 4$ upload slots and 6 queued peers (depicted in white). *On top:* peers are sorted according to a FIFO policy, thus according to the time they entered the queue (i.e., $\tau_1 > \tau_2 > \ldots > \tau_{10}$). *On bottom:* peers are sorted according to the fictitious times calculated by using modifiers (i.e., τ_i^*). Notice younger (in the sense of time spent in queue) but virtuous peers (grey in the figure) enter the first S position, thus starting to receive the content.

5.3. Sharing Enforcers and Bandwidth Caps

When running a file-sharing client interface, a user shares at least three kinds of resources: i) CPU, storage and memory of the hosting machine; ii) files and iii) bandwidth. i) is not commonly a problem, even if a limited storage availability often reflects in a very small amount of file to be served. Notice that such requirements are those strictly necessary to run the client interface. However, for ii) a user may remove a file as soon as it has been downloaded, or enter the overlay without sharing any content, while iii) is crucial both for the reception/delivery of data through a network and to maintain the overlay.

To force user to share files (after all, a file-sharing service without shared files is not so useful), there are two main techniques.

Specifically:

- *automatically share incomplete portion of files:* if a user does not make files available for sharing, then the incomplete portions of currently downloading files are automatically shared. In addition to force non-cooperative users to share data, it is also possible for newcomers to actively participate in data delivery, as soon as a complete chunk has been received. In this perspective, the possibility of handling file chunks instead of monolithic entities (as explained in Section 3.4.1.) is mandatory for the feasibility and efficiency of such a mechanism;

- *exploiting sharing policies in the centralized component:* if a centralized component is present, it is possible to exploit the available knowledge to perform a kind of Call Admission Control (CAC) policy for peers requesting to log-in. For instance, it is possible to deny the access to the overlay to peers not sharing at least x MBytes of data, or y files.

An example of file-sharing architecture relying on a policy in the centralized component is the Direct Connect [44] file-sharing application, where users are coordinated by centralized components called *hubs*. If a user wants to join the network, he/she must comply with the policies enforced by the hub. Notice that sharing policies could also be "qualitative" rather than "quantitative"[4].

Besides, also BitTorrent trackers are now exploiting sharing policy by tracking the *share ratio* of their users. Such policies will be discussed in detail in Chapter 7.

Another interesting technique, implemented in many client interfaces is often called *bandwidth cap*. This mechanism encourages users to set the client's parameters to certain values. A detailed discussion about setting and analyzing users' behaviors when configuring parameters in their client interfaces is available in [45].

A bandwidth cap is a policy that rules the download bandwidth according to the bandwidth set for the upload. Let B_{up} and B_{down} the bandwidth assigned by an user for the upload and download traffic, respectively. Then, a bandwidth cap is a function behaving as follows:

$$B_{down} = f_{cap}(B_{up})$$

Then, $f_{cap}(\cdot)$ constraints the possibility of setting a given value for the B_{down}, according to the assigned B_{up}. The $f_{cap}(\cdot)$ can be designed in different ways. A successful one, that is exploited in the eMule client-interface, will be analyzed in Chapter 6.

5.4. Tit-for-Tat

The Tit-for-Tat[5] (TfT) algorithm has become popular owing to its availability within the BitTorrent client interface [46]. Notice that, strictly speaking, the TfT is not part of the BitTorrent's protocol implementation, rather it is a distributed heuristic to provide a kind of resource management across cooperating peers. In fact, the centralized component employed in BitTorrent, i.e. the tracker (as it will be discussed in detail in Chapter 7) does

[4]For instance, a hub may allow peers to join the system only if they share at least 100 songs and 15 movies, rather than considering only the rough volume of the shared content. This methodology also allows to find out interesting hubs according to their active policy (e.g., a music trader will join a hub forcing peers to share music, rather than movies), hence promoting the self-organization of peers towards common interests or attitudes.

[5]Tit-for-tat is another way to express the concept of *equivalent retaliation*, in the English saying.

not perform any kind of resource management nor allocation. However, in Chapter 8 an optimized framework allowing the tracker to exploit active decisions will be introduced.

The TfT has been proposed as a solution for the *iterated prisoner's dilemma* problem (that will be discussed later in Section 5.4.2.). In fact BitTorrent models the process of allowing/denying a file upload in such a manner.

Specifically, a given peer receives from remote peers different requests for many different file chunks. Then, the client interface, in order to make decisions about satisfying or not such requests employs the TfT strategy. The BitTorrent specification defines the temporary refusal of an upload as *chocking*; notice that even if a peer has been chocked by its counterpart, a download can still take place. In this perspective, delivering data to a peer previously chocked, is defined in BitTorrent as *unchocking*.

Summing up, the TfT strategy has been introduced within client interfaces to elect which peers upload chunks that can assure the highest download rates. In the standard BitTorrent implementation, the number of peers that can be served concurrently is equal to 4. A detailed analysis of the behavior of TfT in BitTorrent systems is available in [47].

Before discussing the iterated prisoner's dilemma, let us briefly showcase the *prisoner's dilemma* game. Then we will briefly analyze the solution provided by the TfT algorithm.

5.4.1. The Prisoner's Dilemma

The prisoner's dilemma is a complete information game originally proposed by Merrill Flood and Melvin Dresher in 1950 [48], and subsequently formalized by Tucker [49], [50]. It is a classical problem of game theory.

We report here the formulation of the prisoner's dilemma game [51]:

Two suspects are arrested by the police. The police have insufficient evidence for a conviction, and, having separated both prisoners, visit each of them to offer the same deal: if one testifies for the prosecution against the other and the other remains silent, the betrayer goes free and the silent accomplice receives the full 10-year sentence. If both remain silent, both prisoners are sentenced to only six months in jail for a minor charge. If each betrays the other, each receives a five-year sentence. Each prisoner must make the choice of whether to betray the other or to remain silent. Each one is assured that the other would not know about the betrayal before the end of the investigation. How should the prisoners act?

The game can be represented by using a bimatrix form, as depicted in Table 5.1.

Table 5.1. Bimatrix representation of the prisoner's dilemma. A and B are the two prisoners (the players), while in parentheses payoffs for each player are reported. Notice that according to the description, payoffs have been reported in years.

	B is Silent	B Betrays
A is Silent	$(0.5, 0.5)$	$(10, 0)$
A Betrays	$(0, 10)$	$(5, 5)$

Let us assume that: i) both players prefer to remain in jail the smallest possible amount of time, and ii) a player does not receive any "utility" for reducing the other one's sentence. In this case, by inspecting payoffs in Table 5.1, the best strategy is that both prisoners betrays. In fact, each player will choose the action that reduces his/her amount of time to be spent in jail. However, if ii) decays, and we imagine that the goal is to minimize the overall amount of time spent in jail by both players (e.g., the payoffs to be considered are the sum of each ones presented in Table 5.1), the new best strategy becomes the one where both remain silent.

Actually, BitTorrent relies on a variant of the proposed game, that is the iterated prisoner's dilemma.

5.4.2. The Iterated Prisoner's Dilemma

The iterated prisoner's dilemma consists in two players playing the game presented in Section 5.4.1. in succession, but with the hypothesis of each player having the memory of the last played game. In this perspective, the actions performed have to be taken into account for the subsequent rounds. As an additional hypothesis, the classical TfT strategy assumes that the first move is "cooperative". Basically, the strategy states that *a player responds in kind of the other player previous action*. Then, if we assume that a player initially cooperates, then he/she will respond according to the opponent's previous action. If the opponent cooperates, the player will cooperate too. If not, the player will not. Moreover, the TfT strategy also introduces the concept of "forgiving". Thus, if the opponent does not cooperate, the player will soon forgive, in order to give chances to start again a mutual cooperation strategy.

This is an efficient solution to this game. Therefore, BitTorrent uses the TfT for chocking/unchocking peers, in order to favor peers that provided the higher upload rates.

The mathematical settings are out of the scope of this book, however interested readers could refer to [50].

5.5. Chunk Selection Strategies

Before discussing different strategies employed to request chunks[6], we point out that each file-sharing application handles them in a different manner. In fact, for performance purposes, a chunk is never treated as a "monolithic" entity. Rather, it is subdivided in smaller sub-parts or pieces, according to the specific application. We neglect such internals here, to keep the discussion as general as possible. Interested readers will found more technical details in Chapters 6 and 7, as well as minor additional tweaks to the general strategies discussed here.

The most widely adopted strategies are the *random selection* and the *rarest first*; while random selection has been utilized in other client interfaces, the rarest first has been successfully introduced in BitTorrent [46].

[6]We use the words chunk and piece interchangeably, since at this level of abstraction there is not the risk of confusing chunks with smaller entities used to avoid performance losses, e.g., sub-pieces.

Random selection is usually employed to enhance the availability of a file when acting in a churn-affected environment. In addition, it avoids the presence in the system of multiple copies of a file stuck at the same level of completion.

As to the BitTorrent system, random selection is used to recover to possible performance bottlenecks introduced by the rarest first strategy. In fact, rarest first implies that a peer requests the rarest chunk available among known peers. In this perspective, the rarest piece could be only available in one peer, thus creating a bottleneck at the "rarest piece owner" side. This is why, such algorithms are often mixed. Especially, the random selection strategy is used in the very initial moment of a new file distribution.

5.6. Peers' Age

As already said, the effect of the churn impacts on the network, both in terms of topological properties and files availability. An interesting technique to mitigate churn effects, is to take some decisions according to the expected life-times of peers populating the overlay.

The original Gnutella protocol specification, and the Gnutella-based Limewire [52] client interface, promote peers to the rank of ultrapeers[7] according to consideration on connectivity (e.g., the presence of a mediating device in the peer access network) and uptime statistics. In fact, as discussed in Section 2.3.2., ultrapeers are expected to have a particular role in the overlay, thus carefully selecting candidates is a plus, also to face failures problem as documented in Section 2.3.3..

Concerning ultrapeer election, it has been noticed that the probability that a client interface remains connected is directly related to how long it has been connected to the network. As a consequence, candidate peers should have a reasonably high uptime to become an ultrapeer [53].

However, there is a lot of research undergoing about this topic. Measuring and quantifying such characteristics is quite a new field. Thus, at the time of writing, age-based algorithms are promising, but it is still hard to quantify their effectiveness in real world deployments. Relevant works on this topic are available in [54], [55], [56] and [57] and in the references therein.

[7] An ultrapeer is the equivalent of a superpeer in the Gnutella nomenclature, as already discussed in Chapter 2.

Part II

APPLICATIONS

Chapter 6

The eMule System

With the foundation provided by the previous chapters, we now introduce the eMule file-sharing application, that is very popular, developed on a continuous basis and released under an open source license. In addition, eMule implements many sophisticated techniques that deserve to be discussed in order to understand the effectiveness of modern client interfaces. Differing from other works about file-sharing applications, we "specialize" here the generic concepts previously explained to enhance their comprehension. Besides, analyzing real-world applications after a general discussion, clarifies how to use the presented general frameworks to describe almost every file-sharing service.

Then, this chapter showcases the design choices, the components and the algorithms at the core of this popular application.

6.1. A Brief Introduction to eMule

The original eMule[1] file-sharing software was introduced in mid 2002 as an enhancement of the original eDonkey2000 client interface[2]. A very talented user, nicknaming himself *Merkur*, was not satisfied of the behavior of the original implementation, so he developed a clone and put it publicly available under an open source license. Still today, the project is hosted on the popular SourceForge website[3], and it has also a dedicated official one[4].

Owing to the availability of the source code, it is possible to reach the soul of the application, and then understand all the protocol internals. However, reconstructing some mechanisms from the analysis of the sources could be time consuming. In this perspective, the most notable effort of documenting the eMule protocol has been done by Kulback and Bickson [58], which offers a detailed discussion about the eMule protocol specification as well as some interesting internals of the client interface.

During the last years (from 2002 to nowadays), many features have been added to the original eMule client interface, and now it could be considered as a totally new system. Among the others, we cite several extensions to the original eD2k protocol, allowing to

[1] At the time of writing, the currently available eMule version is v0.49a.

[2] Notice the pun between eMule and eDonkey.

[3] `http://www.sourceforge.net`

[4] `http://www.emule-project.net`

perform also sophisticated interactions among client interfaces in a fully distributed way. Besides, from the the original "one man" effort, now eMule can rely on an entire team of very skilled developers.

As said, being the eMule source code freely available, many "independent" client interfaces and tweaked implementations, called *mods* (an informal way to say modifications) now populate the scene. Actually, two main kinds of mods exist: i) ports of the original, x86 and Windows-based eMule client interface to other hardware and software platforms[5] and ii) add-ons to the standard client interface, both in terms of "local" features and functionalities. A detailed analysis of the most popular mods, and specific tweaks to the eMule's core algorithms, will be be presented in Chapter 8.

6.2. Supported Overlay Networks

The original eMule system relied on servers for locating users and contents, thus implementing a hybrid p2p architecture (as discussed in Section 2.4.). Servers are essentially the same used by eDonkey2000, and their networked system constitutes the eD2k network.

From v0.40, eMule can also rely on a "serverless" infrastructure, owing to the possibility of organizing eMule clients in a Kademlia [59], [60] overlay network. The specific implementation of the Kademlia protocol within eMule client interfaces has been called Kad. Basically, Kademlia, originally developed by Maymounkov and Mazières is very similar to Chord [21] (presented in Section 2.5.), since it shares with the latter the common technological foundation, i.e., DHT.

In order to fix some inefficiencies of other DHT-based overlays, Kademlia relies on a XOR-based metric to route data more efficiently through the overlay. A detailed comparison among Kademlia and other DHT-based overlays is out of the scope of this work, but interested readers could find more details in [59]. Besides, a deep discussion about Kademlia will be provided later in Section 6.3.. In this perspective, eMule now also relies on a structured p2p architecture (as explained in Section 2.5.).

To augment its effectiveness in locating users and resources, it is possible to use both networks simultaneously, or according to specific needs, only one overlay at time could be accessed. Figure 6.1 depicts the different kind of overlays exploited by the eMule file-sharing application.

6.3. Kademlia

As said, Kademlia[6] is a structured p2p overlay based on DHT. While Chord uses finger table to accelerate the look-up, Kademlia introduces the concept of XOR metric to handle distances and to increase the efficiency of look-up procedures. Besides, like Chord, Kademlia allows to perform searches by contacting up to $O(\log(N))$ peers [59], where N

[5]We cite the most popular one, called aMule, which has been ported over a relevant amount of *nix platforms. The aMule source code is available at the URL `http://www.amule.org`.

[6]In this description, we focus on the specific implementation of the Kademlia protocol within the eMule client interface. Such implementation is called Kad. In this perspective, the terms Kademlia and Kad are used interchangeably.

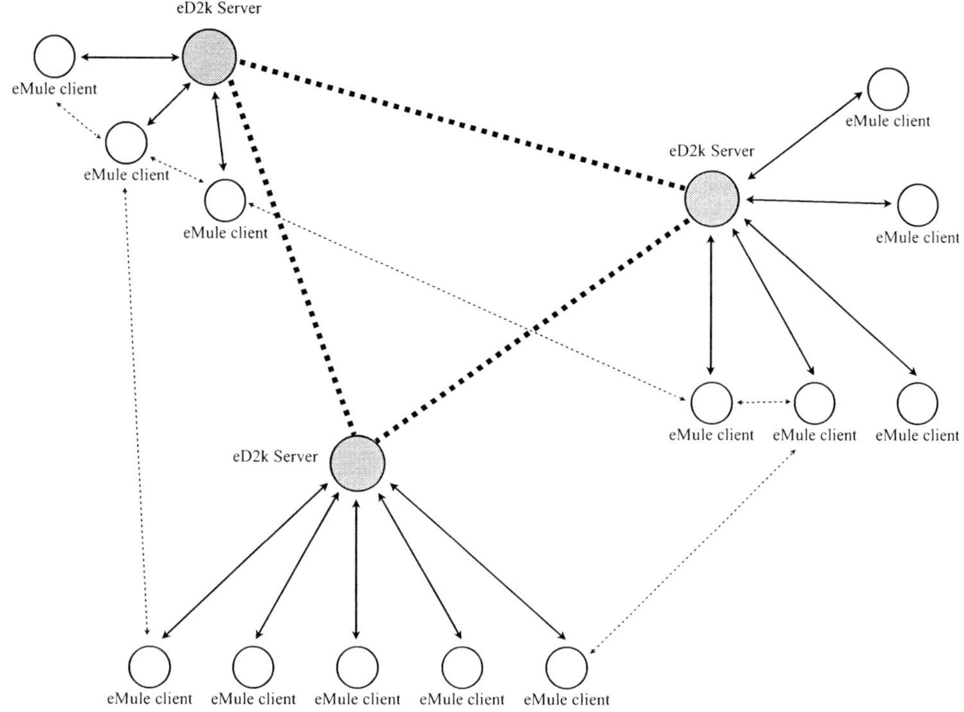

Figure 6.1. The two flavors of overlays available in the eMule file-sharing network. The hybrid overlay and the internetwork of servers. With a grey dashed line are depicted some interaction of the serverless structured Kad overlay network. The p2p connections among client interfaces exchanging data are not depicted.

is the number of peers populating the overlay. Each peer has an NodeID and the same is for files having a FileID. The FileID could be, for instance, the file hash as described in Section 6.7.. Then, like other DHT architectures, it handles $(key, value)$ pairs. To evaluate the "distance" between two peers, Kademlia uses a value, which is obtained by calculating the XOR of two NodeIDs and considering the result as an integer.

The XOR metric has been selected since it is symmetric and obeys to the triangular property[7]. Both NodeID and FileID have the same structure, so it is also possible to apply the concept of distance also when calculating which node has to handle what key. The single routing algorithm implemented in Kademlia permits to threat nodes as leaves in a binary tree. The position of each node within the tree is determined by the shortest unique prefix of its NodeID.

To exploit routing, each peer maintains a list for every bit composing the NodeID space. Then, for a 160 bit wide NodeID, a peer will maintain 160 lists. A list is composed of the data needed to locate another node within the overlay, followed by information such as the IP address, the port number and a NodeID. Since lists have significance for each bit composing the NodeID space, each list corresponds to a specific distance, and nodes in the

[7]For the case of XOR, the triangular property is as follows. For any $a, b \geq 0$, $a + b \geq a \oplus b$.

n-th list must have the same NodeID save for the n-th bit of the NodeID of the node storing the list. Kademlia completes lists with the concept of k-bucket. The value k is a system-wide parameter that defines the degree of replication of routing information; for instance, it is mostly assumed equal to 20. In a nutshell, a k-bucket is a list storing up to k entries.

Concerning the protocol at the basis of Kad, it is composed by four messages carried by UDP. The basic messages are as follows:

- `PING`: this message is used to verify if a given node is still present in the overlay;
- `STORE`: it is used to store a (*key, value*) pair in a given node;
- `FIND NODE`: the node receiving this request will send back the k nodes in his/her k-bucket closest to the requested key;
- `FIND VALUE`: it behaves similarly to `FIND NODE`, but if the receiver has the requested *key*, he/she will send back the corresponding *value*.

Lastly, the Kad implementation within the eMule client interface uses a special file called `nodes.dat` to perform the bootstrap. Such a file stores nodes that are supposed to be present in the overlay. This file must be sometimes updated or downloaded if all the nodes appear off-line or if it is the first time that a client tries to connect to the Kad overlay.

6.4. Operative Modes and File Handling

Concerning the operative modes, an eMule client interface can work in two different flavors: HighID and LowID, which identify a client able or unable to receive inbound connections, respectively. As a consequence, eMule can seamlessly work also when deployed behind a NAT or a firewall. However, when acting in LowID, some limitations apply: for instance, two LowID-ed clients cannot connect to each other[8].

As an evolution of the original eDonkey 2000 client interface, the eMule protocol supports a *multiple download* system, and each peer sharing a given file is called a *source*. Then, an eMule user can concurrently download the same file from different remote peers. In addition, eMule implements also a *hording* mechanism, allowing to exchange information about file sources, in order to contact and locate a larger amount of peers. Typically, this is done via a particular extension of the original eD2k protocol, where data is usually carried out via UDP packets.

To handle the concurrent download of a file, each one is subdivided into chunks of 9.28 Mbytes. Then, a given file is uniquely identified by computing the hash of the concatenation of the hashes of each chunk composing it (this will be presented in detail in Section 6.7.). Moreover, to enhance the distribution and the availability of a content within the eMule network, chunks to download are randomly selected.

[8]Actually, a server-assisted procedure to ensure connectivity among two LowID-ed client interfaces exists; in fact, eD2k servers could also be employed to route data across two LowID-ed clients. However, such feature is almost never enabled, due to the impact on the bandwidth needed at the server side. Notice that, when in presence of two LowID-ed clients, the connection reversal technique discussed in Section 4.3. does not work. For a detailed discussion about the most adopted traversal techniques, refer to Chapter 4 and the references therein.

6.5. Enforcers, Bandwidth Cap and Modifiers

The widespread adoption of eMule is also due to its ability of effectively coping with "freeriders", i.e., users consuming a service without an adequate degree of contribution. In addition, eMule resource management operates in a completely user-independent flavor: for instance, a user cannot overrule decisions taken by the software, e.g., which peer will be served next. Thus, eMule introduces a credit system to quantify and reward users contributing to the network.

Furthermore, the eMule client interface forces the users to share files (at least, if a user does not make files available for sharing, the incomplete portions of currently downloading files are automatically shared) and to share bandwidth.

Specifically, the client interface enforces its available bandwidth, in the sense of how many resources have been assigned to the application. Let us then "specialize" the bandwidth cap function, $f_{cap}(B_{up})$ introduced in Section 5.3..

Recall that the bandwidth cap is used to enforce a relation among the bandwidth available for downloading data, i.e., the B_{down}, and the one for uploading data, i.e., the B_{up}.

Thus, the client interface enforces a relation of the kind $B_{down} = f_{cap}(B_{up})$. Specifically, in the case of eMule, the bandwidth cap enforces the maximum amount of bandwidth that could be set by the user for downloading data. We define such quantity $\overline{B}_{down}$, and the following constraints hold:

$$0 \leq \overline{B}_{down} \leq B_{down}$$

We point out that the following quantities are intended in kbyte/s. Therefore, the bandwidth allocation within a given eMule client interface is enforced through a function $f_{cap}^{eMule}(B_{up})$, designed as follows:

$$f_{cap}^{eMule}(B_{up}) = \begin{cases} \text{if } 0 \leq B_{up} < 3 \rightarrow B_{down} = 3 \cdot B_{up} \\ \text{if } 3 \leq B_{up} < 9 \rightarrow B_{down} = 4 \cdot B_{up} \\ \text{if } B_{up} \geq 9 \rightarrow B_{down} = \infty \end{cases}$$

where B_{up} and B_{down} are the maximum bandwidth limits imposed by the client interface for the upload and download traffic, respectively.

As a consequence of the limited amount of remote peers that can be served concurrently, the B_{up} is subdivided in K equal slots, and the first K peers receive the requested chunk with a rate $\leq \frac{B_{up}}{K}$. Then, each peer serves up to K remote peers, and exceeding peers are queued. Besides, the credit system is used to influence the position within a peer's remote queue, as discussed in Sections 5.2.1. and 5.2.2., respectively.

Depending on the requestor's degree of participation in the file exchanges, an user can achieve a better or worse place in peers' remote queues. In essence, the credit system provides *credit modifiers* of the time spent waiting in remote queues. They are computed by taking into account the amount of data exchanged between a pair of peers. As a consequence, the more a user uploads to a peer, the better is his/her place in the peer's remote queue (the real value of the waiting time is multiplied by the modifier).

As we already performed for the bandwidth cap, let us also specialize the $m(t)$, presented in Section 5.2.2., with the one actually employed in the eMule client interface. The standard eMule client interface calculates two different candidate credit modifiers for each contacted source[9]:

$$R_1(t) = \frac{2 \cdot Up_{tot}(t)}{Down_{tot}(t)} \qquad R_2(t) = \sqrt{Up_{tot}(t) + 2}$$

where $Up_{tot}(t)$ and $Down_{tot}(t)$ are intended as the amount of Mbytes that a peer uploaded to or downloaded from that source (up to the time t of calculation).

Then, $R_1(t)$ and $R_2(t)$ are compared, and a modifier at time t is selected, according to the following relation:

$$m(t) = \min\{R_1(t), R_2(t)\}$$

The modifier $m(t)$ given by the previous relation is subject to some boundaries, imposed to exploit rewards rather than punishment disciplines, and to offer some chances to peers, which have never received data.

Specifically:

- *acknowledging the time spent in the queue*: if a peer has an $Up_{tot}(t) < 1$, then the modifier $m(t)$ is set equal to 1, to prevent penalizing peers that are spending time in the queue;
- *momentary speed increase*: if a peer has never received data (i.e., it has a value of $Down_{tot}(t) = 0$), $m(t)$ is set equal to 10, in order to promote a possible cooperative data exchange;
- *avoiding punishment*: in order to avoid punishment, thus exploiting cooperation through incentives, the lowest possible value for $m(t)$ is 1;
- *preventing "lifetime benefits"*: if a peer is a great uploader, it will achieve huge values for the modifier $m(t)$. However, higher values of $m(t)$ might bring some peers to be "untouchable" for long time, hence causing unfair behaviors. Then, the highest alloted value for $m(t)$ is 10.

We point out again, that the aforementioned $m(t)$ is a specific "incarnation", of the general one presented in Section 5.2.2..

Credit modifiers are then permanently stored, in order to keep track of the contribution of an eMule peer over time, hence not limiting their availability within a single user session. To accomplish that, eMule peers are uniquely identified by a user hash $Hash_{user}$(called also User ID in the eMule terminology as it will discussed in detail in Section 6.7.), which is permanently generated and, to avoid misuse or eavesdropping, the identification between two clients is secured by performing a public-private key exchange.

[9]Again, to keep consistency with the rest of the book, and to avoid burdening of notation, we dropped the dependance from the specific peer.

6.6. Practical Queue Handling

Handling peers in a queue could be resource consuming, both in terms of computing power and bandwidth. In fact, due to churn, user-generated (both locally and remotely) file requests, and evolution of the exchanged volumes, remote peers must be polled periodically to maintain the queue in a consistent state. This implies that the client interface has a constantly running portion of code (e.g., a thread) that reflects changes and updates of peers populating its own queue. In case of long queues (e.g., when in presence of a number of peers $\geq 1,000$), the resulting overheads could be critical, both in terms of computing power needed by the machine hosting the client interface and of bandwidth required to convey the signaling. In addition, remote peers contact a source[10] several times, for instance to ask for other chunks as soon as the pending one is complete.

To cope with such drawbacks, in the standard eMule client interface, the time gap between two subsequent download requests to the same peer, that is called *re-ask time*, T_{reask}, is constrained to $T_{reask} \geq 21\ min$. Moreover, the client interface offers the possibility of setting the maximum number of peers in the waiting queue, but it is also possible to set the number of peers in queue to unlimited. To avoid burdening of requests for different files from the same peer, as well as to not increase the level of complexity in the management of queue modifiers, a peer could be in the waiting queue of a given source only for one file at once. In case a remote peer has more than one needed file, only a request is processed. To increase the flexibility of the process, a peer could ask to swap sources from a file to another, also in an automatic way, in order to change the file for which he/she has been waiting in queue.

As regards the algorithm utilized to sort peers in queue according to their degree of contribution, the actual implementation of the mechanism within the eMule client interface differs slightly from the one presented earlier. The general concept presented in Section 5.2.2. is still valid, but actually instead of defining peers' position according to fictitious time (previously defined, for the i-th peer, as τ_i^*) a special numerical value, called *rank* is utilized. Again, we drop the dependance from i, that indicates that the value is related to the i-th peer in the queue, to avoid burdening the notation. The modifiers are those presented in Section 6.5., and for each peer the rank value, $Rank(t)$, at the time t of computation is calculated as follows:

$$Rank(t) = \frac{m(t) * \tau(t)}{100}$$

we remark that $\tau(t)$ is the time spent in queue by a generic peer up to the time t of calculation. Then, according to the number K of available upload slots, the first K peers with the highest $Rank(t)$ values are served.

To enhance procedures devoted to the entering/leaving remote queues, there are supporting algorithms to promptly report if a contacted peer has an infinite or a full queue. In addition, to avoid long waiting times, a peer could decide to accept to be remotely queued only if its position will be less or equal to for a given amount. In this way it is possible to

[10]We recall here that, according to the eMule terminology, a source is a peer owning the required content.

prevent inefficiencies in the queuing management and enhancing the overall process, i.e., avoiding CPU and bandwidth trashing. In Chapter 8 some modifications of the standard queue management adopted by eMule will be discussed.

6.6.1. Slot Allocation and Friends Management

The eMule client interface allows user to manually set the *slot allocation*, that is the amount of bandwidth to be assigned to each TCP connection responsible of uploading data to a remote peer, defined here as B_{slot}. Therefore, instead of assigning the number of slot K, an "indirect" assignation is done, since the following holds:

$$K = \left\lfloor \frac{B_{upload}}{B_{slot}} \right\rfloor$$

However, users manually assigning bandwidth could bring to a waste of resources due to non-optimal settings. Owing to the cooperative nature of such systems, an inefficient resource exploitation within a client interface, multiplied by the huge number of peers, can reflect in a relevant resource wastage. Then, it is a wise engineering option enabling the client interface to detect non-optimal bandwidth assignment. In this perspective, the eMule implementation is able to recognize if, after distributing the available B_{upload} over the K slots, some bandwidth remains unused.

Let us explain the employed algorithm with an example. Suppose that, $B_{upload} = 15$, and $B_{slot} = 2$, both expressed in kbyte/s. Thus, the number of slots are $K = 7$, but still 1 kbyte/s of bandwidth remains unallocated, since it is not enough to fill a slot with a capacity equal to 2. To cope with this, eMule forces the allocation to have an amount of slots with equal capacity and without trashing, then it forces a $B^{*}_{slot} = 1.875$, corresponding to $K = 8$ slots of equal capacity. This has been also done in order to assure fairness while serving different remote peers concurrently.

In addition, to give users another degree of freedom, it is possible to partially "overrule" the credit system by manually allocating a *friend slot*. This function allows to reserve a download slot for a user that is recognized as a friend. But, to prevent a user from assigning too many slots to friends (which happens when a peer has a number of friends $\geq K$) reflecting in totally ignoring the scheduling computed upon taking into account credit system, only one slot out of the available K could be bound to a friend at a given time.

6.6.2. File Priorities

Until now, remote peers are served according to the mechanisms discussed in Sections 6.6. and 6.6.1.. So, it could appear that the unique scheduling policy is performed according to users' behavior, neglecting the importance of files to be exchanged. Even if the eMule credit system is one of the most advanced and effective [68], the techniques introduced so far do not suffice to assure a satisfactory behavior in different situations.

As an example, let us consider the "release" of a new content. Imagine a user who wants to inject a brand new file on the overlay. In order to exploit the bandwidth available from remote peers, it is important that the new content will be replicated as fast as possible. Consequently, the faster peers interested in the new file will get it (or its subparts), the higher the pool of resources dedicated to its diffusion will become (such concept has been called service capacity, and it will be discussed in Section 7.6.).

To promote the diffusion of new files, as well as to cope with different file availabilities, the eMule client interface introduces the concept of *file priority*. In essence, files can be associated to different levels of priority, in order to interfere with the dynamics introduced by the queue management. To accomplish this, peers in the queue are still scheduled according to time modifiers, but the fictitious time τ^* is weighted also by the "importance" of the file.

Therefore, files with different priorities influence the $Rank(t)$ according to:

$$Rank(t) = \frac{m(t) * \tau(t) * Pri^j}{100}$$

where, Pri^j is the priority of the j-th shared file (or requested file, depending on the particular "viewpoint"). Notice, again, that we drop the dependance of the single peer to avoid burdening notation. Lastly, we highlight that eMule client interface allows to set values for the Pri^j in the range $1,8 \div 0,2$.

6.7. Identifying Resources

In the eMule file-sharing application there are three different classes of key actors that have to be identified: *users*, *client interfaces* and *files*. Each one must be uniquely and precisely identifiable in order to recognize resources, and to maintain both the credit and queuing systems effective.

Let us analyze how eMule handles such identities.

- *user:* as said, a user must be recognizable across sessions[11] in order to track its behavior via the aforementioned credit system. To cope with such a requirement, eMule relies on a randomly generated 128 bit long ID, created by concatenating random numbers;

- *client interface:* actually, both the LowID and HighID (introduced in Section 6.4.) are special kinds of Client ID, which is the identifier used to recognize the specific client interfaces;

- *file:*

  - *file hash:* it is a 128 bit long ID and, as explained in Section 3.4.1., it is calculated by concatenating the hash of each chunk composing it;
  - *root hash:* it is an hash calculated by using the SHA1 algorithm [61] for each file chunk. The algorithm is fed by using data blocks of 128 Kbytes.

Concerning the calculation of Client IDs, eMule adopts the following simple algorithm. Client IDs are server-generated and assigned to the client interface after the initial handshake procedure.

[11] We intend as session, the period of time between when a peer enters and leaves the overlay.

Let a, b, c and d be the octets composing a generic IPv4 address, in big endian format, of a host running the eMule client interface. Thus, $IP_{host} = a.b.c.d$. A tentative ID^* is created according to the following formula:

$$ID^* = a + 2^8 b + 2^{16} c + 2^{24} d$$

Then, if $ID^* > (0x1000000)$[12] eMule assumes the ID^* as an HighID, thus confirming the aforementioned ID^* as the HighID. Conversely, if the $ID^* \leq (0x1000000)$, the server assumes that the client interface has a LowID. The server calculates a different ID, then discarding the tentative ID^*. Even if the client interface source code is publicly available, the server implementation is not. Therefore, it is not known exactly how the LowID is calculated. It has been noticed that LowIDs differ from server to server (even if the IP_{host} does not vary), and this hints at some kind of random generated procedure. Notice that the presented formula guarantees that HighIDs do not vary from server to server or from session to session (as long as the IP_{host} remains unchanged).

[12] 16777216 in decimal notation.

Chapter 7

The BitTorrent System

As it has been done for the eMule system, we describe the BitTorrent file-sharing application by relying on the foundation provided by the previous chapters. Such application is very appreciated and widely adopted owing to its simplicity and high efficiency in exploiting the available bandwidth.

Therefore, similarly to Chapter 6, this chapter showcases the design choices, the components and the algorithms available within this file-sharing application.

7.1. Concepts and Architecture

BitTorrent has been developed in 2002 in order to cope with scalability problems of centralized content distribution systems. The original developer was Bram Cohen. Briefly, it is not a classical file-sharing system, since its purpose is to solely replicate content over a networked environment in an efficient manner. As a consequence, both the protocol and the client interface are "lightweight".

This has been possible owing to clear engineering choices, specifically:

- the architectural components are well defined and designed to efficiently solve simple duties;
- there is a clear separation between the cooperative role of peers and coordinating duties of the centralized component;
- "companion" features, such as content searches, are performed outside the overlay exploited via the BitTorrent protocols;
- some tasks, i.e., the hashing mechanism of content, are performed off-line by a third entity[1].

Prior to analyzing the internals of BitTorrent, we investigate the architectural blueprint and we describe its components. Figure 7.1 depicts the basic overall architecture composing the BitTorrent system.

[1] Even if this "third entity" may not be physically separated from the client interface, it has to be considered as a standalone entity, at least conceptually.

Specifically:

- *Distribution Swarm*: a group of peers participating in the exchange of a given content. A peer could participate in distributing different files simultaneously. However, for each file a different swarm is established;
- *Tracker*: the centralized component responsible of creating and supervising the swarm (e.g., providing user paging facilities);
- *Web-Server*: is not mandatory, but it has been depicted in the reference scenario to highlight the concept that the `.torrent` file has to be retrieved "outside" the BitTorrent architecture; commonly, it is available through web pages, hence the presence of the web-server in this reference example;
- *Peer*: is a user who wants to obtain the seeded file. Actually, there are at least two kinds of peers, specifically: i) seeder, which is a peer owning the entire content, thus acting as a server and ii) peers with an incomplete copy of the file, thus sending/receiving information.

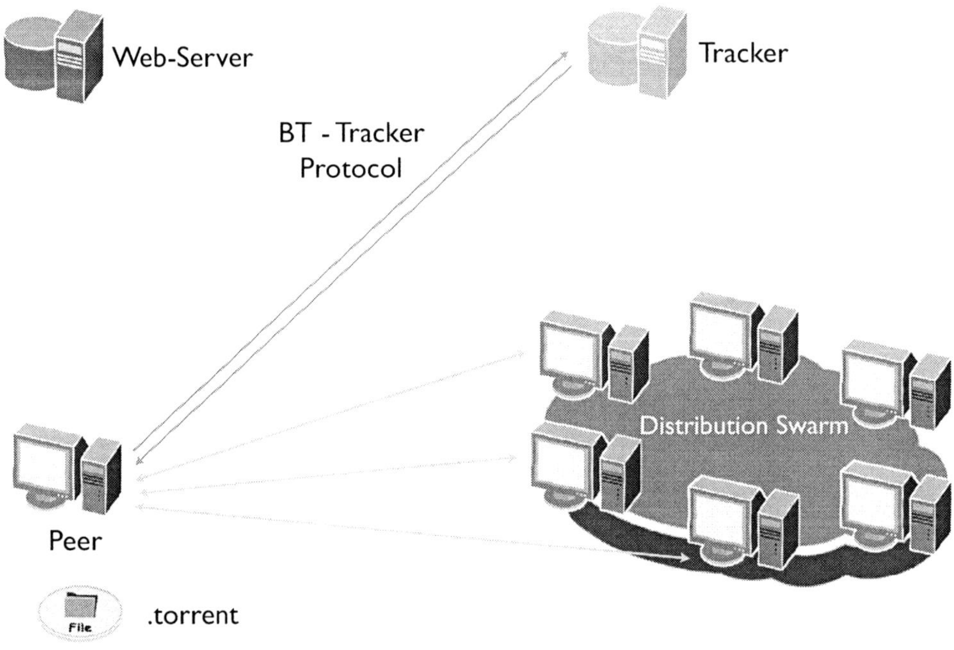

Figure 7.1. The overall BitTorrent architecture. Notice the presence of the BT - Tracker Protocol, employed by the client interfaces for interacting with the centralized component. Therefore, this allows to completely decouple the p2p portion from the rest of the system.

As it happens in other p2p file-sharing applications, also here the content to be seeded, which is a generic piece of information, is subdivided in to smaller parts called *chunks*.

In order to speed-up transfers, different chunks can be retrieved by different peers concurrently. However, a given chunk must be retrieved by only one peer at time, and the same chunk could be delivered to different peers simultaneously. In addition, even if not completely available in one peer, a complete replica of the content to be disseminated could be available in the swarm. In fact, the totality of chunks composing the content could be replicated in a distributed fashion, being available in different remote peers. BitTorrent highly relies on this technique, which is called a *distributed copy*.

We point out that, in the literature, the parts composing the content are not always denoted by the same term. In fact, many Authors are used to refer them as a pieces, chunks, blocks, ... In the BitTorrent jargon, chunks composing a file are widely called *pieces*, and they typically have a size of 512 Kbytes. Other sizes exist and the most common are: 256 Kbytes and 1 Mbytes. Notice that varying the size of a piece does not only reflect in a better/worse utilization of the available transmission resources. In fact, pieces are also "described" in the .torrent metadata file (as it will be showcased in Section 7.1.2.), thus accounting for its bigger/smaller dimension. However, since a file is quite always not a perfect multiplier of the aforementioned quantities, the last piece will have an irregular size.

Then, the number of pieces composing the content, defined as M is given by:

$$M = \left\lceil \frac{FileSize}{PieceSize} \right\rceil$$

where, $FileSize$ and $PieceSize$ are the dimension of the overall file and the dimension of the piece composing it, respectively.

Lastly, BitTorrent has been designed to be able to efficiently handle *flashcrowds*, which are sudden peaks of simultaneous requests [62], [63]. This dramatic scalability accounted for its rapid diffusion; more details about how this is achieved are presented in Section 7.6..

7.1.1. Life-Cycle of a Torrent and Peculiarities

It is worth to better investigate the differences in the life-cycle of files shared with "classical" p2p file-sharing systems (e.g., Gnutella and eMule) and those made available through BitTorrent.

In "classical" systems, the client interface makes files shared by an user available for the entire time the software is running. Files can be dynamically shared/unshared by acting over preferences. Conversely, in BitTorrent, files are not permanently shared (in the "classical" sense). Therefore, the typical life-cycle of a torrent (i.e., a file seeded through this particular architecture) is:

1. a user wanting to inject a content prepares a .torrent, which has to be handled by a tracker. Then, the .torrent has to be delivered to interested peers. Such user then becomes the first seeder, i.e., a peer owning the entire content and, at the very beginning, the only onc;

2. upon obtaining the .torrent, a peer exploits the log-in procedures with the tracker. Then, the phase of data exchange within the swarm takes place;

3. as soon as a peer finishes the download of a file, it becomes a seeder, continuing to deliver data to remote peers until the client interface is stopped.

Until here, the life-cycle is quite similar to those of other systems, even if the .torrent shifts issues such as content searches outside the particular file-sharing overlay. However, some major differences are as follows:

1. the first implementation of the client interface, being very simple, discouraged to resume seeding torrents. Thus, in order to maintain the torrent belonging to a content active, the client interface had to be kept running indefinitely;

2. even if many participants are active (also solely acting as seeders), a given torrent exists if and only if the tracker remains active. Losing the tracker responsible of a torrent accounts for the file unavailability. Of course, different .torrents, handled by different trackers could coordinate different swarms to reduce such hazards. However, the swarms belonging to different trackers are physically decoupled and, from a logical point of view, the content, even if the same will be handled as different files to be seeded;

3. every .torrent "points" to a given tracker. Changing the tracker reflects in preparing another .torrent and delivering it again[2].

To cope with such issues, modern client interfaces handle automatically the resume of files. Files are managed as standard shared entities, and the logic within the client interface to perform all the needed operations to revamp a given torrent is completely transparent from the user. Also, as it already happens in eMule, client interfaces implementing the BitTorrent protocol specification are also providing support for structured p2p overlays, thus bypassing the rigidity of the centralized and single tracker-based design.

7.1.2. Structure of a .torrent

The .torrent file is defined by its creator as a *metainfo file*. It is a typical example of metadata, i.e., it contains data about other data. As said, it allows to shift complexity about data search outside the overall architectural blueprint. Such files are generated by using ad-hoc utilities or by using "wizards" available within the more sophisticated client interfaces. This design also decouples the preparation of the content to be disseminated from the content delivery architecture itself. For this reason, we describe here its structure, since it is simple yet effective.

Data within the .torrent file is encoded by using a scheme, called *bencode*[3]. Bencode supports the four following types: byte strings, integers, lists and dictionaries. Details about the bencode are outside the scope of the book, but interested readers will found more internals here [64]. We point out that the original BitTorrent specification does not specify the maximum number of bits that have to be used to represent integer. However, to be able

[2]This could not happen if the .torrent contains a list of backup trackers, which have been specified prior to preparing this metainfo file. For more details, refer to Section 7.1.2..

[3]It has to be pronounced *"Bee Encode"*.

to describe quantities greater than 4 Gbyte, support for signed 64 bit integer is mandatory, thus reflecting in the capacity of the .torrent to support file greater than 4 Gbyte.

There are two kinds of .torrent files, namely: i) metadata files describing torrents composed by a *single* file and ii) metadata files describing torrents composed by *multiple* files. Both species share the following logical "blocks":

- **info**: it is a dictionary describing file (or files) available in the given torrent;
- **announce**: it contains the URL of the tracker responsible of the given torrent;
- **announce-list**: it is an optional field and contains a list of backup trackers. Notice that it is an extension to the official BitTorrent specification;
- **creation date**: it is an optional field containing the creation date of the .torrent in standard Unix epoch format[4];
- **comment**: it is an optional field and embodies textual comments;
- **created by**: it is also an optional field and presents the name of the author and the software that created the .torrent.

The dictionary named **info** exists in two variants, depending on whether the .torrent describes single or multiple file(s) torrents. It is composed of a common part and one which changes according to the specific flavor. The common part is as follows:

- **piece length**: it is an integer containing the number of bytes composing each piece;
- **pieces**: it is a string representing the concatenation of the 20 byte hashes of each piece, which has been computed via the SHA1 algorithm;
- **private**: it is an optional field declaring if the torrent has to be handled as private, i.e., no external peers are allowed to act as sources.

Then, according to the flavor of the .torrent, the **info** dictionary contains different information. For the single file case:

- **name**: it is the name of the file;
- **length**: it is the length of the file in bytes[5];
- **MD5 checksum**: it is an optional field containing a 32 bit long checksum of the file computed via the MD5 algorithm.

Otherwise, for the multiple file case:

- **name**: it is an "advisory" file name of the directory where to store files;

[4]Unix epoch (also called Unix time) describes the time via the number of seconds elapsed since the midnight of January 1st 1970 in Coordinated Universal Time (UTC).

[5]Notice that, as said in the beginning of this section, the need of using bencoded integers sufficiently "big" is to avoid getting impeded in handling large files.

- **files**: it is a list of dictionaries describing each file composing the torrent. Each entry of this dictionary has the following form:
  - **length**: it is the length of the file in bytes;
  - **MD5 checksum**: it is an optional field containing the MD5 checksum, as previously explained;
  - **path**: it is a list containing strings describing the path of each file.

Notice that files are still described as in the previous case, but the field **path** has been introduced to place all the files in a correct location within the filesystem.

7.2. Tracker Duties

As briefly hinted at in Section 7.1., the tracker is responsible of supervising a set of torrents, composed by disjoint swarms, also providing some kind of basic coordination among peers. Particularly, the role of the tracker is to inform a peer interested in participating in the download of a given file of other peers already populating the swarm. This is performed delivering a list of peers (by using a proper protocol, as it will explained in Section 7.3.) allowing a peer to directly exchange data by using the p2p portion of the protocol (as it will explained in Section 7.4.).

Concerning the list of peers, typically it contains 50 entries (of course, if the swarm is composed of less than 50 entities, the list will be smaller) and peers to be included in the list are selected randomly by the tracker. A client interface is also able to ask for a larger peer list by specifying the amount of peers via the parameter **numwant**, when interacting with the tracker. We point out that is also possible to better engineer and optimize the composition of the list. In Section 8.2.2., a discussion about the optimization of BitTorrent-like content delivery strategies will be introduced, with a particular attention to the exploitation of the tracker.

Commonly, a given peer has a list composed by 80 remote peers. This fact, jointly with the random flavor of the selection algorithm, reflects into a swarm built on an overlay completely ignoring locality of nodes. In other words, clusterization of peers composing the swarm rarely happens. On one hand, this increases the robustness of the overlay with respect to underlying network failures or congestions, while, on the other hand, it reflects in a less efficient usage of resources and increasing delays (e.g., data is transferred among geographically remote peers though an overloaded international backbone). The aforementioned remarks should be "evaluated" keeping in mind concepts explained in Section 2.2.1..

Lastly, trackers are also employed to exploit policies to force cooperation and to assure a proper contribution within a given community. Such mechanism will be discussed in Section 7.4.1..

7.3. BitTorrent Tracker Protocol

To perform communications among client interfaces and the tracker, a simple protocol built on top of the HTTP and HTTPS standards is employed. From the viewpoint of a BitTorrent peer, the tracker is a minimal web server (e.g., a kind of webservice) answering to HTTP GET requests. The URL is composed by the base URL defined in the announce field within .torrent, as described in Section 7.1.2. and additional parameters are "appended" by using the Common Gateway Interface (CGI) model, thus:

$$\texttt{http}:\backslash\backslash\texttt{www.thetrackerurl.com}\backslash?\texttt{param1}=\texttt{value1}\&\ldots\&\texttt{paramN}=\texttt{valueN}$$

A full description of the parameters is out of the scope of the book: interested readers should refer to the protocol specification available in [65].

Summing up, there are two kinds of communications: *tracker request parameters* and *tracker response*. As regards the tracker request parameters, the exchanged values are used to identify a peer, the port on which the client is listening on, the true IP address of the machine running the client interface, its "state" in the sense of exchanged bytes, and its operative status, i.e., started, stopped and running.

The tracker answers to client interfaces by sending back tracker responses, which are *text/plain* documents containing a bencoded dictionary with different keys. Again, the complete discussion is out of the scope of this work, but the major ones are as follows:

- **interval**: it represents the time in seconds a client interface should wait before sending new requests to the tracker. Yet client interfaces could interact with a more aggressive behavior (e.g., ignoring the value specified in interval). This happens when peers have something to report (for instance, reporting a forthcoming graceful disconnection from the swarm) or when they ask for another list of peers. Periodically re-asking for the list of peers allows to recover to the churn affecting the swarm, hence preventing that the "perceived" portion of the overall swarm progressively "evaporates". In the latter case, the suggested procedure is to ask for a larger list of peers, instead of *hammering* the tracker with multiple requests. As it happens in other file-sharing applications (e.g., eMule), peers with a too aggressive behavior could be banned[6]. This is mandatory in order to avoid the saturation of the capacity of the tracker;
- **complete**: it contains the number of peers owning the entire content, i.e., seeders;
- **incomplete**: it contains the number of non seeding peers;
- **peers**: it is a dictionary used to "describe" the swarm. Each entry has the following keys:
  - **peer ID**: a 20 byte string used as a unique ID for the client interface;

[6]Peers repeating a request too often could reflect into a kind of DoS or Distributed DoS (DDos). This is why subsequent interactions are often "clocked". Typically, a peer re-asks the list of other peers composing the swarm every 300 s, but this time could be changed by the administrator of the tracker. The mechanism of defining a "waiting time" by using the Interval value, is something conceptually similar to the re-ask time introduced in eMule and explained in Section 6.6.. Notice that, if in presence of controlled settings, both the Interval and the re-ask time could be tweaked to better exploit the available resources. This will be explained for eMule in Section 8.1.2..

– **IP address**: it specifies the address of the peer. It can contain the IPv4 or the IPv6 address, or the DNS name[7];

– **port**: the port number used by the peer.

Besides, an important extension is the one related to handling communications about the *scrape*. The scrape extension allows to receive information about all the torrents a tracker is in charge of managing. The name resembles the operation of "screen scraping", i.e., manually fetching the web pages containing the status of the tracker. Additional details on the scrape extension (also including unofficial extensions) are available here [65].

7.4. The Peer Wire Protocol

The peer wire protocol is responsible of convoying communications among remote peers and to actually deliver chunks composing the content to be seeded. In addition, it relies on the TCP for transport services and it conforms to the original BitTorrent philosophy of maintaining the design simple. The peer wire protocol is then responsible of performing four basic operations:

- handshaking: that is the set of operations needed to successfully establish an overlay connection with a peer participating in the swarm;

- choking/unchoking signaling: that is the set of basic communications needed to coordinate peers continuously playing the TfT to maintain an efficient exploitation of the available resources;

- requesting chunks: such portion of the protocol is responsible of specifying the block needed by a peer. In a nutshell, it carries chunk requests;

- data: the actual movement of data from/to a remote peers.

To maintain this work short and focused we omit here the complete analysis of the peer wire protocol. Interested readers will find more details in [65].

7.4.1. Centralized Heuristics and Community Management

As said, BitTorrent relies on a distributed and game-theoretic based framework to promote cooperations among peers. However, also centralized mechanisms have been deployed. In fact is not uncommon to have trackers organized in sort of "communities", i.e., the website hosting the .torrents also provides trackers and discussion forums. Thus, there is an attempt to maintain the data exchange within a given community of users effective, meanwhile keeping track of virtuous members and purging users exhibiting an extreme leeching behavior.

[7]According to the particular "address flavor", different encoding mechanisms could be used. In fact, the IPv4 address is expressed in the dotted notation, the IPv6 address is represented in the hexadecimal form and the DNS name is encoded as a string.

Then, centralized heuristics have been developed. In particular some communities keep track of the degree of contribution from every users. To do so, a kind of *Global Contribution Ratio* (GCR) is calculated, as follows:

$$GCR = \frac{V_{up}}{V_{down}}$$

where, V_{up} and V_{down} are the traffic volumes, in Mbytes, generated by a peer in its entire lifetime within the given community, for the upload and the download of contents, respectively. Thus, it is possible to perform different heuristics to guarantee (or to force) a proper degree of participation even if the distributed mechanisms appear insufficient. When this centralized mechanism is in place, at least two actions could be performed:

1. Perform a CAC policy by preventing access to the swarm of peers having a $GCR \leq \alpha$, where α is a threshold defined by the tracker administrator;

2. impose to peers having a $GCR \leq \alpha$ a kind of "enforcement cycle", i.e., allow them to enter a swarm only for delivering data in order to re-balance the too aggressive leeching attitude[8].

Relying on such policies rises the problem of how to identify a given peer (or a client interface) to keep track of its contribution and exploiting the actions previously presented. The most common approach is to "personalize" the .torrent file for every user.

The typical procedure is:

- a user must have an account to be identified within the community;

- in order to retrieve a .torrent, the user must be recognized by the system;

- the system applies a digital signature inside the .torrent in order to bind the participation of a user in a given swarm and to keep track of future actions;

- the GCN is updated regularly and restored from session to session.

Thus, when such procedure is exploited is important that a user does not exchange its .torrent file with other users. In fact, the system has the "granularity" of .torrent-level, hence it will update the GCN whenever a peer starts to join the swarm by using that particular .torrent file. Besides, many communities also prohibit (or at least highly discourage) the adoption of a BitTorrent client interface also implementing distributed search heuristics (e.g., by using DHT-based mechanisms to page users, contents and trackers). In fact, it could happen that the client interface exchanges data among peers belonging to different communities (but exchanging the very identical content) and hence it is impossible to precisely calculate the GCN of a given peer.

[8]A typical value for α is 0.25. Interested readers can find a community ruled by the presented heuristics at the URL `http://www.dimeadozen.org`. We are reporting this BitTorrent community, since it is used to exchange authorized "bootlegs", i.e., live recordings of musical performances authorized by the performers themselves, and then completely legal.

7.5. Distributed Heuristics

With the broad term *distributed heuristics* we define a set of mechanisms employed to increase the efficiency of BitTorrent, both in terms of data exchange and fairness. Then, we cite here the TfT (already discussed in Section 5.4.) and the unofficial extensions applied to modern client interfaces (e.g., integration with The Onion Router [66], DHT support,...). Since the latter are very fast-moving, we invite readers to investigate the software directly. A good starting point is to understand the tweaks employed in the Azureus BitTorrent client interface, available at `http://azureus.sourceforge.net`.

We only briefly introduce here the *end game* technique introduced in the original BitTorrent implementation used to enhance the efficiency of the replication process.

The end game is basically a variant of the standard heuristics employed to request chunks, and it is triggered when a download of a file is almost complete[9]. Notice that when employing a supervised mechanism as discussed in Section 8.2.2., such a feature is inessential. When only a small amount of pieces are needed, there is a tendency to receive them slowly. Basically, this is due to the reduced level of "concurrency" that can be gained by requesting several pieces in parallel. To mitigate this fact, a client interface acting in end game mode sends the request for all the missing pieces to all the known peers. To avoid a dramatic waste of resource, as soon as the requested pieces are received a "cancel" message for that piece is sent (by using the peer wire protocol showcased in Section 7.4.) to all the remaining remote peers contacted.

7.6. Service Capacity

We decided to introduce and discuss the concept of service capacity at this point of the book, since the swarm-based, single file, BitTorrent system is well represented by the seminal work of de Veciana and Yang [67]. We do not explain all the concepts presented in this paper, since they are out of the scope of the book. Instead we discuss only the behavior of the service capacity in a deterministic model of the transient phase. In addition, we only present the detailed derivation in the case of monolithic files, i.e., files to be exchanged are not further subdivided in chunks. Anyway, readers are very encouraged to investigate all the concepts proposed by the Authors in the original work. To favor the reading of the original work, we also borrowed the notations employed.

The deterministic model of the transient analysis of the service capacity of p2p file-sharing applications, allows to understand the disruptive power available for content delivery operations. Before starting, we introduce the definition of *service capacity*, according to the one adopted by de Veciana and Yang. Service capacity is the number of peers available to serve the document[10].

[9]The concept "is almost complete" is not well explained in the protocol specification, thus resulting quite obscure. Alas, there are not any documented values when to trigger the end game mode. For instance, some client interfaces switch to end game when all the blocks composing a file have been requested, while others follow different heuristics. Thus, it is also clear the importance of being able to observe the produced traffic patterns and the source code to understand such internals, if not publicly available or well documented.

[10]In our case, the document is the generic content to be seeded, and it has been extensively called file across the entire book.

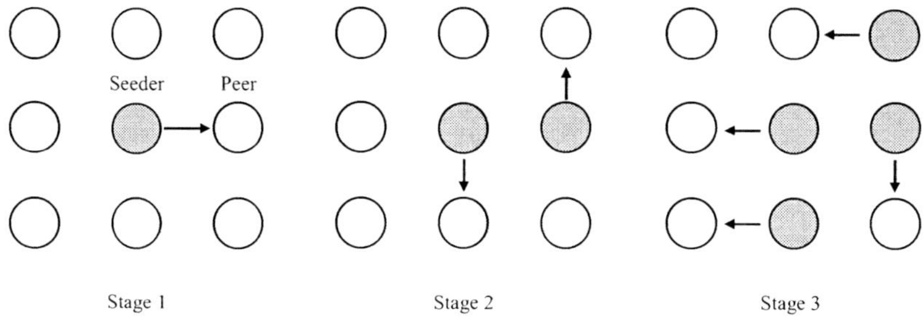

Figure 7.2. Reference scenario employed to illustrate the transient phase of the service capacity of p2p file-sharing applications. A seeder, i.e., a peer owning the entire content, is depicted in gray. Peers requesting data are white. Notice that only the first three stages have been reported. Stage 1 shows the injection of a content within the swarm: only one peer owns the content. Stage 2 and 3 represent the first two evolutions of the system.

Figure 7.2 depicts the reference scenario adopted to analyze the transient phase of the service capacity. In particular, let us suppose that at the *Stage 1*, a seeder owning the entire content wants to start the distribution within the swarm. Besides, let us suppose that the swarm is composed by n peers. Then, $n-1$ peers are interested in acquiring the file originally available only in the seeder. In order to keep calculation simply, let us assume, without loss of generality that $n = 2^k$. Suppose that each peer has an upload capacity of b bit/s and the content to be seeded has a size of s bit. In addition, let us assume that a seeder will serve another peer only when it has delivered the entire content. Lastly, notice that no churn is present, i.e., the number of peers populating the swarm is constant.

Then, to serve $n-1$ peers, $s(n-1)$ bit of data have to be exchanged across the swarm.

Let us consider what is intuitively the *good* strategy of serving *one* peer at the rate of b in order to increase the service capacity of the overall swarm to $2b$. Then it is possible to serve *two* peers, and so on, until the entire population of $n-1$ requesting peers is served entirely. The first two steps of this evolution are depicted in Figure 7.2 and labeled as *Stage 2* and *Stage 3*, respectively. If we adopt such idealized strategy, peers will receive the whole file every $\tau = \frac{s}{b}$ seconds. This reflects on an exponential growth in the available service capacity, following a trend of $2^{\frac{t}{\tau}}$, where t is the time.

Under such assumptions, the $n-1$ peers are served in a time frame equal to $\tau \log_2(n+1)$. Recall that, as hypothesis, we choose $n = 2^k$, then, the time frame becomes equal to τk.

Now, it is interesting to evaluate the average delay, $\bar{d}$, experienced during this period, that Authors define as *transient regime*. Let d_j be the delay experienced by the j-th peer, and note that $2^{i-k}n$ peers receive the file at the time $(i+1)\tau$. Let us assume that the first peer receiving the file has zero delays.

Then:

$$\bar{d} = \frac{1}{n}\sum_{j=1}^{n} d_j = \sum_{i=0}^{k-1} 2^{i-k}\tau(i+1) = k\tau - \frac{n-1}{n}\tau$$

It is possible to further approximate the formula, resulting in:

$$\bar{d} = \tau(\log_2 n - \frac{n-1}{n}) \approx \tau \log_2 n$$

Even if in presence of an initial burst of n concurrent requests, the average delay experienced by a peer $\bar{d}$ scales with a logarithmic trend, i.e., $\log_2 n$, which is better than the linear scaling typical of client-server systems.

Let us consider the case when the content to be seeded is not handled as a monolithic entity. Rather, it is subdivided into m chunks of the same size, according to techniques explained in Section 3.4.1.. In this perspective, the average delay experienced by the j-th peer become influenced by the number of chunks m, thus $d_j^{(m)}$. Then, the average delay is equal to:

$$\bar{d}^{(m)} = \frac{1}{n} \sum_{j=1}^{n} d_j^{(m)} \approx \frac{\tau}{m} \log_2 n$$

This means that the subdivision of the content to be seeded into m chunks, accounts for a reduction of m times the average delay experienced in the transient regime. However, we cannot imagine to increase m too much, since extra signaling and overheads will impact dramatically in the overall performance.

The proposed idealized models give some hints about the opportunity and the strengths of employing the p2p paradigm for user-based file-sharing operations. We also point out the impact of the churn in the service capacity: in fact, a peer abandoning/entering the swarm both accounts for a loss of resources, as well as a reduction of the overall service capacity achievable upon completion of the content.

Part III

Analysis, Research and Internals

Chapter 8

Underground and Research

As said, p2p file-sharing is a phenomenon generated from the "underground", and still major advancements are performed in such an environment. Many researchers have been analyzing behaviors of file-sharing applications for several years, as well as proposing enhancements and new algorithms.

In this perspective, this chapter offers a general portrait about the major developments performed by the Internet Community in the field of p2p file-sharing applications. Among the others, we will investigate some client interface modifications, specific tweaks to support community-wide deployments, and the usage of web cache to enhance file transfers over the file-sharing infrastructure. Besides, we also introduce some solutions developed in the Academic field to enhance such platforms.

The major findings and the notations utilized to describe eMule modifications to support a particular network setting, optimization of p2p file-sharing applications, and some results have been partially borrowed from [68], [69] and [70].

8.1. Underground Developments

As mentioned earlier, the source code availability for about the totality of modern client interfaces spawned two main actions: i) the port of major file-sharing applications on different operating systems and hardware architectures, thus increasing the overall diffusion and user population and ii) the birth of ad-hoc client interfaces for particular network settings. As regards i), the most "ported" application is BitTorrent, owing to its lighteweight protocol specification. Consequently, different client interfaces exist, but the core protocol, as well as the basic behaviors, remain unchanged. Actually, the main efforts in BitTorrent development appear to be focused on user-tracking aware communities, i.e., `.torrent` repositories also capable of delivering personalized metadata and functionalities to keep track of users' contributions, and to empower BitTorrent clients of DHT-like distributed search support (as showcased in Chapter 7). Notice that such goals are often conflicting.

Concerning ii), the most modified client interface is the eMule one. Actually, there are several modifications; ranging from very simple tweaks (for instance adding features devoted to save known sources), to massive modifications to support some particular network settings or community of users operating over a common network infrastructure.

8.1.1. eMule Webcaches

The *eMule webcache* technique consists in encapsulating file chunks inside web traffic and then using ISPs' webcaches to both speed-up the file transfer operation (if enough bandwidth at the cache-side is available) and to reduce the needed bandwidth at the peer side (since chunks are delivered only to the cache and the further redistributed by it). Notice that the exploitation of web caches is a direct consequence of the file handling techniques presented in Section 3.4.2.. In fact, caches are not designed to "trap" big contents (e.g., in the order of 10 Mbytes). Rather they are engineered to deal with small-sized contents, which are the standard in web-based traffic. Thus, relying on smaller sub pieces for transferring a file is mandatory to do the magic.

We point out that this mechanism also produces some beneficial side-effects on the eMule network: for instance leechers (people who download files without uploading anything) will impact less on the overall p2p network. There are also some benefits for people who access the Internet by using not a flat-rate contract, but a pay-per-traffic basis. In fact, several ISPs do not count traffic from their proxy server and web caches in the monthly limit. Nevertheless, also ISPs experience some benefits: caches are usually deployed to provide a load reduction in critical portion of the network.

However, the misuse of the web caches could bring to swamping the caching infrastructure, consequently reducing its effectiveness when handling the original web traffic. It is also true that, in order to effectively benefit from this enhancement, some modifications of the eMule protocol are needed. One of the key points is to keep the credit base system used for quantifying the "sharing attitude" of an eMule user balanced.

One possibility is that the classical credit system remains the same, but with an enhancement, allowing to recognize clients that successfully create a positive-proxy uploading, in order to get rewarded by mean of some extra credits. However, in the current eMule implementation credits are disabled when exchanging traffic with a webcache. Obvioulsy, "webcached" transfers only happen if two peers demand the same file behind the same cache. Lastly, the effectiveness of the proposed approach is highly questionable, owing to cache performance and availability, and it also rises severe security and privacy issues.

8.1.2. Community-Based Enhancements to eMule

In order to show a real-life and user-based engineering effort applied to p2p file-sharing applications, we present the modifications done to the eMule client interface to support a wide area community, which relies on a particular network setting. Specifically, this is the case of a nation-wide eMule community built on the network of a single ISP offering a flat-rate Fiber To The Home (FTTH) connectivity, if available, with a capacity of 10 Mbit/s. If not, a standard Asymmetric Digital Subscriber Loop (ADSL) over copper is deployed. In such case, the maximum capacity is reduced to 6 Mbit/s and 512 kbit/s, for the downstream and the upstream, respectively.

The ISP offers Internet access through a NAT and it basically operates a nation-wide private IP network, which resembles an interconnection of high-speed Metropolitan Area Networks (MANs). To overcome transparency limitations introduced by the exploitation of NAT devices, many solutions are available, as discussed in Chapter 4. However, this particular modification is interesting, since it is based on modifications not aimed at exploiting traversal techniques, but at better employing the possible cooperation among peers living in the same network environment.

In this perspective, this "case study" allows to appreciate the soul of p2p, which is based on a systematic cooperation among peers to overcome hazards, also exploiting the isolation properties available through the usage of a well-suited overlay network.

For coping with the lack of transparency of peers belonging to the ISP-made private realm, the eMule client has been properly modified, to work seamlessly both with private and public IP addresses. In fact, due to the presence of the NAT, every eMule user acting on this ISP is LowID-ed when interacting with a host outside the ISP network. In addition, this penalizes users running eMule within this ISP, since they cannot contact nor connect with other LowID-ed client interfaces, reducing their effective participation in the overall file-sharing system. Conversely, when directly interacting, they all lay before the NAT and hence can establish connections directly, appearing as HighID-ed. More details on how the IDs are handled by the eMule system are available in Section 6.7..

The credit system, and the modifiers remain unchanged (i.e., they remain those explained in Section 5.2.1. and 5.2.2., respectively), but two separate queues have been introduced: this is the one of the most relevant modifications done and it will be discussed in detail in the next section.

As a consequence of the presence of NAT, users belonging to this ISP represent a kind of *split space* in relation to the rest of the world. The main reason is due to the aforementioned LowID-HighID relationship imposed by the eMule protocol, as explained in Section 6.4.. Thus, in order to recover to this "eMule divide" scenario, after a campaign of empirical analyses, developers introduced a *dual queue mechanism*.

Let us analyze in detail its internals. Let us define as Q^C the queue only for users belonging to the community, while Q^{Ext} as the queue for any other eMule peers or unofficial community mods. As a consequence of this design, a peer allowed to enter Q^C will only climb this queue and the same applies for peers in Q^{Ext}.

The dual-queue mechanism gives benefit for both peer classes. For community users, the waiting time is reduced. In fact, being N^C the overall amount of users belonging to the same community, and N^{Ext} the overall amount of external users equal to $\sim 4,000,000$; then $N^C \ll N^{Ext}$.

We point out that the dual-queue architecture reflects in a pseudo-dynamic allocation of the available upload bandwidth, particularly if employed jointly with the mechanism explained in the following Section. Also, notice that, as long as Q^C remains empty, all the service capacity is allocated for Q^{Ext}.

As discussed, the dual queue mechanism allows to prevent a community user from climbing a highly populated queue, thus reducing the waiting time for serving peers belonging to the same community. As another degree of freedom, users can also decide to manually allocate the amount of the overall upload bandwidth to be used for serving peers in Q^C and Q^{Ext}, with the following constraint

$$B^A_{upload} + B^{Ext}_{upload} = B_{upload}$$

where, B^A_{upload}, B^{Ext}_{upload} are the upload bandwidth set in the client interface to deliver data to community and external peers, respectively. In addition, B_{upload} is the total available bandwidth alloted for the upstream traffic.

If the manual bandwidth allocation is not desired, a users can also rely on a *fair enforcement system*, that automatically adjusts B^{A}_{upload} and B^{Ext}_{upload} in order to obey the following constraint:

$$\lim_{t \to \infty} \left\{ \frac{Vol(t)^{External}_{downloaded}}{Vol(t)^{External}_{uploaded}} \right\} = 1$$

where $Vol(t)^{External}_{downloaded}$ and $Vol(t)^{External}_{uploaded}$ are the cumulative data volumes downloaded and uploaded with non-comunity peers, at the time t of calculation, respectively.

The proposed enforcement system tries to guarantee the fairness of the modified eMule with respect of the rest of the world, by allocating an amount of bandwidth for peers outside the community equal to the one received. The fair enforcement system is active iff $Q^{C} \neq \emptyset$, since, to avoid bandwidth trashing, external eMule clients will be served even if the limit of 1 has been hit. As a consequence of this design choice, it ensures a balanced reward to external eMule clients, but it rewards them more than needed, on average.

Another important tweak has been done in the eMule logic devoted to contact sources. As explained in Section 6.6., peers are forced to obey a time constraint, defined as T_{reask}, when asking to download a given content, despite the availability of free download slots. In this modification, the T_{reask} among community users, T^{C}_{reask}, has been constrained as $T^{C}_{reask} \geq 7\ min$, in order to cope with the diffuse availability of high speed accesses.

Conversely, for contacting peers outside the community, it remains unchanged. The reason is that unmodified clients would ban peers asking too often for a chunk, resulting in too aggressive requesting behaviors.

Regarding the network, this eMule modification supports both standard eD2k servers and the Kad network, as it happens in the original implementation. However, in this case, eD2k servers are employed only for locating external eMule clients, while the Kad network is used for only locating clients running inside the ISP. The reason is rooted within the eMule development history. In fact, the first Kademlia implementation used in the standard eMule was unable to support clients under a firewall/NAT (hence, LowID-ed) until version v0.44a[1]; as a consequence, the mod inherited such peculiar behavior. Even if with some limits, the adoption of Kad as "internal locator" has been preferred to an internal eD2k server, owing to its distributed flavor, in order to reduce the risk of legal threats for server maintainers.

File transfers across peers must rely on "self balancing" algorithms, since the client interface could be adopted over different access technologies with different bandwidths. This property is intrinsically guaranteed by employing TCP for data transfer.

But, the peculiarities of the application layer protocol, for instance swarming-based mechanisms and chunk-grained data delivery play a major role in the achievable throughput, despite the availability of bandwidth. This is the case of the setting under investigation, where peers often act over FTTH links, exchanging data with each other, thus having a relevant amount of resources for transmitting files.

[1]This modified eMule version has been originated from the source code of v0.44a of the standard client interface. Then, developers started to implement tweaks on this tree, in order to have a working application. Syncing with more recent developments of the standard client interface is something that happened later.

According to tests on client interfaces, the eMule system appears to reach the maximum throughput of ~ 160 kbyte/s per single connection. Two different aspects account for performance bottlenecks: i) protocol implementation and ii) queue temporization.

Regarding the protocol implementation, as discussed earlier, the eMule protocol relies on packetized data of 180 kbytes. This leads to objective difficulties of managing multiple source downloads at high data rates. A simple way to by-pass such limit, without breaking compatibility with classic client interfaces, could be to overcome the "granularity" of the scheduler employed for managing the download slots (one of the first K positions in the queue).

It has been noticed[2] that the eMule client interface has an option allowing to send only complete chunks. This means that if a client reaches a download slot, it can be removed from that position only when it finishes downloading a complete chunk. Conversely, if this option is not enabled, removal can happen with every sub-parts of 180 kbytes. Basically, this option changes the "granularity" for managing the first K positions of the waiting queue.

In addition, also the re-ask time plays a role. In fact, a peer can ask for a content every 21 minutes in the original implementation, and 7 minutes for peers running the modified version. However, if a peer finishes a transfer in a time frame $\hat{t}$, and $\hat{t} < T_{reask}$, it must wait before starting another transfer for a period of $T_{reask} - \hat{t}$. To mitigate this, the modification implements three strategies: i) the re-ask time has been reduced (as explained in Section 6.6.); ii) the option forcing the transmission of a complete chunk is active by default; iii) the modified client interface (only for community-to-community transfers) sends two complete chunks one after the other, i.e., an amount of data equal to 18.56 Mbytes[3]. We highlight that sending two complete chunks instead of increasing the dimension of a single one allows not to break the hashing mechanism, thus guaranteeing compatibility.

Then, this tweak allows to "fill-the-pipe" and, when in the presence of high-speed connectivity, it guarantees to reach higher transfer rates (per single TCP connection).

8.2. Research Applied to File-Sharing

The research community has been studying p2p file-sharing systems for years. The most relevant amount of works is in the field of traffic analysis and modeling, since their impact over the network is very critical. Nevertheless, characterizing their traffic patterns allows to build synthetic models to evaluate performance via simulative analysis. To keep the discussion structured and simple, we will introduce the state of the art of research on traffic analysis of file-sharing application in Chapter 9, which is dedicated on discussing how to collect and understand traffic patterns.

Major advancements have been done in the field of structured overlays, i.e., the development of Kademlia and DHT-based networks. Such concepts have been discussed in Chapter 2 and 6. Modeling file-sharing systems in a "monolithic" flavor is an hot topic in the research. In this perspective, in Chapter 7 we proposed a brief overview on the seminal

[2]Actually, inspecting users' behaviors in setting preferences in client interfaces is something very valuable to engineer p2p file-sharing applications. Details on how to gather such information and how to exploit them are provided in Chapter 9.

[3]Since, according to the eMule implementation, each chunk is 9.28 Mbytes, we have a net amount of data delivered equal to $(9.28 \cdot 2) = 18.56$ Mbytes.

work of de Veciana and Yang, which is also handy to give a rough quantitative perception of the disruptive power of p2p file-sharing systems, and of the benefit of relying upon chunk-based algorithms.

Another relevant topic and engineering effort concerns *optimization*. Optimization allows to better utilize both peers and network resources, that are critical for the successful exploitation of p2p architectures. Then, we present a self-contained discussion about two different efforts to optimize p2p file-sharing applications. The first one concerns the optimization of the queue management of eMule, while the second one deals with an optimized control strategy for data replication in a BitTorrent-like environment. We decide to present such works since they give an idea about how to model part of a p2p file-sharing application, what are the main issues when optimizing p2p systems, which have peculiarities like churn and often accounts for the curse of dimensionality.

8.2.1. Optimization of the eMule Queue Management

As it will be presented in Chapter 9, the modifier mechanism implemented within the eMule client interfaces suffers of some accuracy and fairness issues. In this perspective, for the purpose of dynamically optimizing the modifiers assigned by a peer's client interface [69], we introduce a *feedback* control scheme. Thus, there is some logic within the client interface to decide the value of the modifiers at the beginning of each time period of given length Δt. Then assume that the node[4] can gain perfect information of the transfer (upload and download) capacity of each connected source. Hints on how achieve this will be discussed in Chapter 10.

Let us define some notation conventions that we utilize through this section; y_i indicates the i-th component of a given vector y. Then, define N^* as the total number of remote peers connected to the node, and N as the number of peers among them that have a content the node wants to download (i.e., they can upload to the node). We will refer to them as *paired peers*. Define b_0^d, b_0^u as the total download and upload bandwidth of the node assigned to the eMule application, respectively, and b_j^d, b_j^u as the download and upload capacity devoted to the transfer of data with the j-th paired peer. The actual upload bandwidth of the node is equal to b_0^u/S for each remote peer, since the eMule client interface allows to serve a maximum of S remote peers simultaneously.

Next, define $\tau_i(t)$, for $i = 1,\ldots,N^*$, as the total amount of time a remote peer has been connected to the node at time t. Finally, for the j-th paired peer, $j = 1,\ldots,N$, define the total amount of Mbytes uploaded to the node and the the total amount of Mbytes dowloaded from the node at time t as $u_j(t)$ and $d_j(t)$, respectively.

Then, the state vector $x(t) \in \mathbb{R}^{N^*+2N}$ at time t is defined as

$$x(t) = [\tau_1(t),\ldots,\tau_{N^*}(t),u_1(t),\ldots,u_N(t),d_1(t),\ldots,d_N(t)]^T$$

This allows to model the scenario depicted in Figure 5.1. Next, define the control vector $m(t) \in \mathbb{R}^N$ as the vector of the modifiers for the paired peers, i.e., $m(t) = [m_1(t),\ldots,m_N(t)]^T$, and furthemore let (without loss of generality), for $i = 1,\ldots,N^*$, $m_i^*(t) = m_i(t)$ if $i \leq N$ and $m_i^*(t) = 1$ if $N+1 \leq i \leq N^*$.

[4] In the rest of this section, we use the term *node* to identify the peer under investigation.

Also consider, for $j = 1 \ldots, N$, the discrete random variable $\gamma_j(t) \in \{0,1\}$, which is equal to 1 with probability $P_j^\gamma[x(t)]$ and 0 with probability $1 - P_j^\gamma[x(t)]$, and take $\gamma(t) = [\gamma_1(t), \ldots, \gamma_N(t)]^T$.

With these definitions, we can write a state equation of the form $x(t+1) = f[x(t), m(t), \gamma(t)]$ in the following way

$$\begin{aligned} \tau_i(t+1) &= \tau_i(t) + \Delta t, & i = 1 \ldots, N^* \\ u_j(t+1) &= u_j(t) + \min\left\{ \frac{b_0^d}{N^\gamma}\Delta t, b_j^u \Delta t \right\} \gamma_j(t), & j = 1, \ldots, N \\ d_j(t+1) &= d_j(t) + \min\left\{ \frac{b_0^u}{S}\Delta t, b_j^d \Delta t \right\} \delta_j[m(t)], & j = 1, \ldots, N \end{aligned}$$

where N^γ is the number of paired peers that upload to the node and $\delta_j[m(t)] \in \{0,1\}$ is a function of the control vector computed as follows. Define $\tau^*(t) \in \mathbb{R}^{N^*}$ as the vector of the elements of the set $\{\tau_1(t)m_1^*(t), \ldots, \tau_{N^*}(t)m_{N^*}^*(t)\}$ sorted in descending order (i.e., $\tau_1^*(t) \geq \tau_2^*(t) \geq \ldots \geq \tau_{N^*}^*(t)$). Then we have $\delta_j[m(t)] = 1$ if $\tau_j(t)m_j(t) \geq \tau_S^*(t)$ and $\delta_j[m(t)] = 0$ otherwise.

This quantity determines whether a paired peer can be among the S peers who download from the node after the modifier has been applied.

We also define $\lambda_j(t) = \dfrac{u_j(t)}{d_j(t)}$ as the ratio between the amount of data upload and downloaded by the j-th peer.

As regards $\gamma_j(t)$, we model through this quantity the probability, $P_j^\gamma(t)$, that a paired peer is actually uploading to the node during stage t. Such probability obviously depends on how much the peer has uploaded to the node with respect to how much it has downloaded[5].

In this way, when the node is uploading to the remote peer much more than it is downloading (i.e., $\lambda_j(t) \ll 1$), the node has a higher probability of remaining in the first positions of that peer's queue. Whereas, as the ratio $\lambda_j(t)$ gets higher, it is more likely that other peers will be given higher priority, due to their higher modifiers.

For the purpose of the optimization of the transfer, we define a cost functional related to how "balanced" the data flow through the node is at stage t.

In particular, define $\bar{\lambda}(t)$ as $\dfrac{1}{N}\sum_{j=1}^{N} \lambda_j(t)$ and consider the following cost

$$h[x(t), m(t), \gamma(t)] = \left|\bar{\lambda}(t+1) - 1\right| = \left| \frac{1}{N}\sum_{j-1}^{N} \lambda_j(t+1) - 1 \right| \tag{8.1}$$

Such cost is the average of the upload/download ratio with respect to all the paired

[5]To present this optimized approach, we model the probability $P_j^\gamma(t)$ through a sigmoidal function of this kind:
$P_j^\gamma(t) = 1 \quad \frac{1}{1+e^{-\beta(\lambda_j(t)-\alpha)}}, P_j^\gamma(t)$, where $\alpha, \beta \in \mathbb{R}$. Such method has been selected to keep the simulation model simple and easy to implement. In fact, in real world, precise information about peers' ranks is easily achievable by using the standard eMule protocol.

In addition, in order to feed the predictive control, some probability distribution must be estimated, for instance by evaluating the actual rank of paired peers.

peers. Then, if h is close to one, the exchanged volume of data among the paired peers is averagely balanced, i.e., the node is acting "fairly".

As we want to optimize the system through a horizon of stages, we adopt a *model predictive control* technique [71], by which we look online for a sequence of controls that minimizes the cost over the next K stages. Then, we apply only the first control vector to reach the next stage, and repeat the procedure for the new state. Concerning the number of "prediction stages" K, we can expect that larger values of K lead to more effective strategies. Yet, larger values of K correspond to a heavier computational burden, thus we need to consider a tradeoff between effectiveness and computational effort.

As regards strategies for the online optimization over the K stages, we adopt a scheme that aims at finding the control sequence that minimizes the average of the cost (estimated through an empirical average over a finite number of randomly extracted sequences) over the K stages. In particular, we want to solve the following problem: find

$$\min_{m(t),m(t+1),\ldots\ldots,m(t+K)} \sum_{k=0}^{K} \sum_{q=1}^{Q} \frac{1}{Q} h[x(t+k), m(t+k), \gamma^{(q)}(t+k)] \tag{8.2}$$

subject to $1 \leq m(t+k) \leq 10$ for $k = 0,\ldots,K$, where $\{\gamma^{(q)}(t+k)\}_{k=0}^{K}$, for $q = 1,\ldots,Q$, is a sequence of vectors randomly extracted according to the probabilities P^{γ}. The cost $\sum_{k=0}^{K}\sum_{q=1}^{Q}\frac{1}{Q}h[x(t+k),m(t+k),\gamma^{(q)}(t+k)]$ for a given sequence $m(t),m(t+1),\ldots\ldots,m(t+K)$ and a given initial point $x(t)$ can be computed online by means of simulation.

Regarding the minimization, we notice that the cost is piecewise-flat with respect to the control, due to the fact that different values of the modifiers produce different effects only if the corresponding peers enter or leave the set of the S ones that can download during the stage. Thus, standard descent search methods are not feasible, as the gradient is null.

To cope with such cost we employ a Monte Carlo[6] procedure [72], which in our context consists in sampling the control space with L points, choosing them in such a way that no regions of the space are neglected (which would be unreasonable in absence of prior information), i.e., extracting them with a uniform probability. Then, we evaluate the cost corresponding to such points, and the point corresponding to the lowest value of the cost is taken as the solution.

As said, after the optimal sequence of controls $m^{\circ}(t), m^{\circ}(t+1),\ldots\ldots, m^{\circ}(t+K)$ has been derived, the first control vector $m^{\circ}(t)$ is applied to obtain the actual new state $x(t+1)$, which depends on the actual online value of $\gamma(t)$, and then the procedure is repeated. We remark that the resulting control is in closed-loop form, i.e., it depends on the current state $x(t)$.

Notice also that we only optimize the transfer corresponding to paired peers, leaving the modifiers of peers that are not paired always equal to 1 by default (consistently to typical eMule systems).

In order to evaluate the effectiveness of the proposed optimized queue management, we conducted a performance evaluation campaign via simulations. Here we present only one use case, but interested readers could refer to [73] and [68] for a more detailed performance analysis. Specifically, we investigated the behavior of the optimized system when

[6]Notice that quasi-Monte Carlo sequences can also be employed instead of purely i.i.d. sequences, for faster convergence [74].

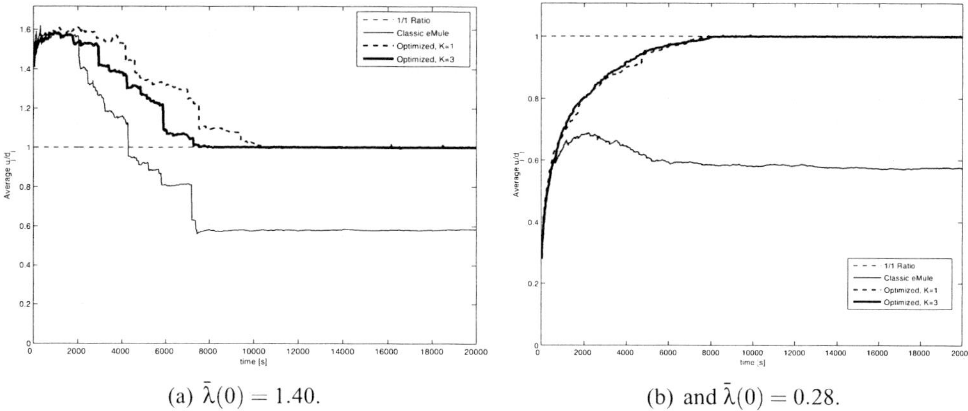

(a) $\bar{\lambda}(0) = 1.40$.

(b) and $\bar{\lambda}(0) = 0.28$.

Figure 8.1. Plots of the u_j/d_j ratio trend of different strategies employed to handle modifiers when the node acts on an asymmetrical access.

an eMule client interface acts over an asymmetrical access, with a downstream capacity $b^d = 1024$ kbit/s, and a downstream one $b^u = 640$ kbit/s. To better investigate the impact of the asymmetrical access over the enforcement strategies, we analyzed both selfish and altruistic initial behaviors (i.e., $\bar{\lambda}(0) > 1$ and $\bar{\lambda}(0) < 1$, respectively). As regards the number of paired peers N is 100 and the total number of queued peers N^* is 1000. The number S of available upload slots is equal to 4.

The initial waiting time $\tau_i(0)$ for each peer populating the queue was taken as a random variable with uniform distribution in the 0-100 seconds range. For each paired peer, the initial amount of uploaded and downloaded data with the node were chosen in order to reflect the characteristic of the scenario, in terms of the initial ratio $\bar{\lambda}(0)$. The upload and download capacities of queued peers assigned to the transfers with the node have been taken in the range 3-30 kbit/s per slot and 16-128 kbit/s, respectively. Such values have been chosen to reflect heterogenous access characteristics (e.g., from dial-up to high-speed Internet accesses).

Figures 8.1(a) and 8.1(b) depict the best trends of $\bar{\lambda}(t)$, when the node has an asymmetrical bandwidth provisioning and when different initial values of $\bar{\lambda}(0)$ are utilized. Specifically, when $\bar{\lambda}(0) = 1.40$ the fixed heuristic implemented by eMule tends to force the node to reach the *1:1* ratio. However, the system turns out to be unfair, since the eMule strategy forces the node to lock on an asymptote corresponding to a $\bar{\lambda}^* \simeq 0.6$.

Conversely, when employing the proposed optimized strategy, with $K = 3$, after a transient period of about 2 hours, the node reaches the *1:1* ratio and it maintains such a value, reflecting in an actual fair treatment. Notice that, when in presence of huge up-time behaviors, thus having the transient period negligible, it is possible to use the optimized strategy with $K = 1$. In such case, some extra time to reach the *1:1* ratio is needed, but a lower computational effort is required. In Figure 8.1(b), the same set-up with a specular initial behavior (i.e., $\bar{\lambda}(0) = 0.28$) is analyzed. As shown, the eMule system is still unfair: after an initial phase when the credit system tries to recover the "data deficit", the node stabilizes its trend around the ratio of 0.6, reflecting in an unfair policy. On the contrary, the optimized strategies reach the *1:1* ratio. Summing up, it appears that the behavior of the eMule system could be endangered by the asymmetrical nature of the access on which the client interface operates. In fact, statically assigned modifiers do not provide enough flexibility to adapt the "progression" according to effective bandwidth allocations.

8.2.2. Optimization of BitTorrent-like Strategies

In this section we discuss how to exploit the centralized component utilized in BitTorrent (presented in Chapter 7), i.e., the Tracker, to actively coordinate peers participating into a given distribution swarm [70]. In addition, we will give ideas in Chapter 10, how this method could be effectively implemented in a real client interface. Before introducing the analytical model developed to exploit optimization, we characterize the peer, according to the perception of the tracker itself.

Therefore, we can model the i-th peer as a bundle of resources identified by the set $\{b_i^u, b_i^d, I_i(t)\}$, where b_i^u and b_i^d are the upstream and downstream bandwidths available for the peer, respectively, and $I_i(t)$ is the vector describing the level of completion of a given content seeded via its swarm (each component is representative of a particular chunk composing the overall content). In particular, for each peer i, $i = 1, \ldots, N$, define

$$I_i(t) = [c_{i,1}(t), \ldots, c_{i,K}(t)]$$

where $c_{i,k}(t) \leq M$ (expressed in Mbytes) is the level of completion of chunk k at time t for peer i, and K is the number of chunks composing the content to be distributed. Both the bandwidths are assumed to remain constant over time, i.e., a peer is not subject to other competing flows generated by other applications. However, we will remove such hypothesis in the following, in order to better reflect realistic scenarios. We define a peer for which $c_{i,k}(t) = M$ for all k and all t as a *seeder*.

In order to optimize the performance of the system, we employ a discrete-time dynamic model to describe the exchange of chunks among peers through a given time horizon, as well as a control scheme similar to the one presented in Section 8.2.1.. This lets the Tracker react to the actual evolution of the swarming process. To this purpose, we need to assume that the tracker has always perfect information of the state and of the bandwidths of each peer. Furthermore, it takes notice of arrivals and departures at the beginning of each time period.

Thus, we need to assume the discretization interval Δt is sufficently small and the churn of peers are "slow" enough to guarantee that both the number and configuration of peers remain basically constant during the whole interval.

The state vector at time t is

$$x(t) = [I_1(t), \ldots, I_N(t)]^T \in \mathbb{R}^{NK}$$

Define, for peers i, j and chunk k, the term $y_{i,k,j}(t) \in [0, 1]$ as the bandwidth share devoted to the upload of chunk k from peer i to peer j. Notice that terms of the kind $y_{i,k,i}(t)$ are obviously equal to 0.

Also define $\delta_{i,k,j}(t) \in \{0, 1\}$ as $\delta_{i,k,j}(t) = 1$ if chunk k is uploaded from peer i to peer j at time t, 0 otherwise.

For what concerns the state equation, we have that the evolution of the size of the chunks from stage t to $t+1$ follows, for $i = 1, \ldots, N$ and $k = 1, \ldots, K$

$$c_{i,k}(t+1) = c_{i,k}(t) + \sum_{j \neq i} \delta_{j,k,i}(t) y_{j,k,i}(t) b_j^u(t) \Delta t \tag{8.3}$$

For positive integers $P \leq N-1, i \leq N$, define a subset of indexes $\{j_1, \ldots, j_P\} \subset \{1, \ldots, N\} \setminus \{i\}$. An *exchanging loop within peers* in the network for a given chunk k is represented by the following equation

$$\delta_{i,k,j_1} + \sum_{p=2}^{P} \delta_{j_{p-1},k,j_p} + \delta_{j_P,k,i} = P+1$$

We say that a network within a distribution swarm is feasible at stage t if it obeys the following rules:

1. peer i can download a given chunk k from one peer only;
2. peer i can not transfer chunks to itself;
3. peer i can upload no more than S chunks at the same time;
4. the network must not contain exchange loops within peers in order to be admissible.

In order to obtain a "compact" description of the swarming process, for the aims of optimization it is convenient to reformulate the problem in terms of the S maximum connections that define a feasible network within a distribution swarm.

To this purpose define, for peer i, $R_i(t)$ as the set of indexes corresponding to peers that download at least one chunk from peer i during time stage t, i.e.,

$$R_i(t) = \{j \text{ such that } \delta_{i,k,j} = 1 \text{ for some } k\}$$

and

$$r_i(t) = [r_{i,1}(t), \ldots, r_{i,S}(t)]$$

where $r_{i,s}(t) \in R_i(t) \cup \{0\}$. We have $r_{i,s}(t) = 0$ if $R_i(t) = \emptyset$. In such case, peer i is not uploading.

Then, define

$$h_i(t) = [h_{i,1}(t), \ldots, h_{i,S}(t)]$$

where $h_{i,s}(t) \in \{1, \ldots, K\}$, as the chunk associated to connection s at time t. We introduce the pair $\Theta_i(t) = [r_{i,s}(t), h_{i,s}(t)]$ to denote which chunk the peer i is uploading to which peer using connection s. Then, we lump the two members of the above defined pair into a vector $\Theta_+(t) = [r_1(t), h_1(t), \ldots, r_N(t), h_N(t)]$ which specifies completely the structure of the network.

Furthermore, define

$$q_i(t) = [q_{i,1}(t), \ldots, q_{i,S}(t)]$$

where $q_{i,s}(t) \in [0,1]$ is the upload bandwidth share corresponding to connection s, and

$$q_+(t) = [q_i(t), \ldots, q_N(t)]$$

With these new definitions, the new control vector at time t becomes

$$u(t) = [\Theta_+(t), q_+(t)]^T \in \{1, \ldots, N\}^{NS} \times \{1, \ldots, K\}^{NS} \times \mathbb{R}^{NS} \tag{8.4}$$

It is also convenient to reformulate the state equation in a way that is equivalent to (8.3), but better suited to the definitions introduced in this section and to the optimization procedure described in the next section.

In particular, for every peer $i = 1, \ldots, N$ and every chunk $k = 1, \ldots, K$, we have

$$c_{i,k}(t+1) = c_{i,k}(t) + q_{j_{i,k}, s_{i,k}}(t) b^u_{j_{i,k}}(t) \Delta t \tag{8.5}$$

if there is a pair $[j_{i,k}, s_{i,k}]$ with $j_{i,k} \in \{1, \ldots, N\} \setminus \{i\}$ and $1 \leq s_{i,k} \leq S$ such that $h_{j_{i,k}, s_{i,k}} = k$ and $r_{j_{i,k}, s_{i,k}} = i$ (i.e., peer i is downloading chunk k from peer $j_{i,k}$) and

$$c_{i,k}(t+1) = c_{i,k}(t)$$

otherwise.

In order to optimize the overall performance of the system, suitable cost functions have to be minimized *online* by the the tracker (i.e., the decision-maker). Thus, we need to specify indexes of the efficiency of the content replication process as the scenario evolves.

Due to the churn affecting the system, we need to maximize the benefit that can be achieved in the transition from stage t to stage $t+1$, i.e., we need to look for cost functions that are well suited to a so-called *myopic* strategy.

A cost function that can be used is as follow: define $c_i^{tot}(t) = \sum_{k=1}^{K} c_{i,k}(t)$ and $c^{tot}(t) = [c_1^{tot}(t), \ldots, c_N^{tot}(t)]$. Then, we have

$$J(x(t), u(t)) = \|c^{tot}(t+1) - MK \cdot 1_N\|$$

where 1_N is a vector made up of N 1's. Such cost represents the target that every peer must have all the chunks as soon as possible. For other possible costs to be employed, interested readers could refer to [70]. We point out that, owing to the "freedom" available in designing the cost function, we successfully tested such a framework over vehicular networks. Interested readers can find more details in [75].

Then, we consider a hybrid random/nonlinear programming scheme that aims at finding optimal solutions to the problem in a fast and computationally feasible way.

In particular, we propose a two-phase procedure that consists in (i) randomly selecting an admissible network configuration and (ii) performing a real optimization over such network. Specifically, the algorithm at stage t can be described as follows.

1. Update the number of peers N according to the arrivals and departures at the end of stage $t-1$. Compute the new state vector $x(t)$ accordingly.

2. For $l = 1, \ldots, L$, select a network configuration that is admissible according to the previously defined constraints, i.e., find a random set $\{\Theta^1, \ldots, \Theta^L\}$ where each $\Theta^l \in A$.

3. For any $l = 1,\ldots,L$, define $q_+(t) = [q_i(t),\ldots,q_N(t)]$ and consider the following real optimization problem

$$\min_{q_+ \in [0,1]^N} J(x(t),\Theta^l(t),q_+(t))$$

Define $q_+^{*,l}(t)$ as the argument of such minimization.

4. Compute

$$J^*(x(t)) = \min_{l \in \{1,\ldots,L\}} J(x(t),\Theta^l(t),q_+^{*,l}(t)) \tag{8.6}$$

and define $\Theta^*(t) = \Theta^{\hat{l}}(t)$ and $q_+^*(t) = q_+^{*,\hat{l}}(t)$ where $\hat{l}$ is the index that minimizes the cost in (8.6).

5. Consider $u^* = [q_+^*(t),\Theta^*(t)]$ as the (sub)optimal control for stage t, and apply it to the system to obtain $x(t+1)$.

With respect to the computational effort, we have to perform L real optimizations in a SN-dimensional space, which is actually deployable in an on-line context with large values of N and K and small Δt. The algorithm turns out to be fast, and easily implementable due to the many real optimization packages available for different software and hardware platforms. Besides, owing to the source code availability of the tracker, we can also imagine a seamless"merge" of such packages and real systems.

Of course, we can expect to obtain better solutions as the number L of "sampled" networks grows, at the expense of more computational time. Concerning point 2) of the algorithm, various strategies can be adopted to obtain a randomly selected feasible network, as explained in [70].

The optimization method has been compared to common strategies that implement the typical actual behavior of unoptimized p2p systems, including BitTorrent. Specifically, at each stage t a feasible network Θ is chosen, and then the upload rates $q_i(t)$ are assigned according to the following rules.

1. The total upload bandwidth b_i^u is shared equally among the S^* active connections, i.e., connections s for which $r_{i,s}(t) > 0$. Specifically, we have $q_{i,s} = b_i^u/S^*$, $s = 1,\ldots,S^*$, $S^* \leq S$.

2. If the total downloaded bytes of peer i for all the chunks exceed the maximum capacity $b_i^d \Delta t$, the upload rates of peers uploading chunks to peer i are scaled in such a way that the sum is equal to b_i^d, while mutually maintaining the original proportions in terms of uploaded bytes.

3. When needed, the upload rates are further corrected, so that each peer can transfer, for chunk k, no more than what is needed to bring the other peer to the same level of completion of that chunk at the beginning of stage t.

As regards the choice of the feasible network, we considered two strategies: (i) random selection as in the case of the optimized algorithm and (ii) *rarest first* strategy[7].

We point out that rule 1) takes into account the presence of a TCP-like transport layer, which somewhat guarantees competing flows to share quite equally the available bandwidth [76]. For what concerns 3) such bandwidth assignment is introduced to make the "fair" algorithm more similar to the way existing unoptimized real p2p systems work. Actually, in general the constraints are even stronger, given peers cannot upload a chunk until they download it completely.

In highly heterogenous environments, congestion issues may happen at any time during the replication process, due to which we must obviously expect some degradation of the performance. Therefore we introduce now congestion phenomena, as to reflect realistic use cases. Actually, we employ here the term *congestion* to define two different reductions of the bandwidth available for each peer to establish a remote connection.

Specifically:

- *network congestion*: it is representative of a situation where the underlying network infrastructure is not able to transport all the offered load. In this perspective, peers with an established end-to-end connection may not be able to transmit data at the maximum available rate, due to presence of bottlenecks representing congested links (or nodes);

- *local congestion*: it is representative of congestion in the *local link* used by each peer to access the network. Local congestion can be due to the presence of other applications generating competing flows, hence reducing the bandwidth available.

Both kinds of congestion involve one or more of the S connections of a peer, with the effect of decreasing the actual amount of data transfered through that connection.

We can model the effect of congestion by introducing a quantity $\gamma_{i,s} \in [0,1]$ defined as the percentage of transfer loss due to congestion on connection s of peer i, and by modifying (8.5) in the following way

$$c_{i,k}(t+1) = c_{i,k}(t) + (1 - \gamma_{j_{i,k},s_{i,k}}) q_{j_{i,k},s_{i,k}}(t) b^{u}_{j_{i,k}}(t) \Delta t$$

Even if there exist mechanisms for detecting the presence of congestion, in general it is not possible to model its impact on the transfer of data nor its evolution, therefore the tracker must solve the optimization problem as if no congestion is present during the interval of length Δt. Of course, the actual presence of congestion might lead to real tranfers at the end of the stage that are less than what is expected in the optimization phase. Nevertheless, the closed-loop structure of the controller makes the tracker able to react to the new state.

To evaluate the effectiveness of the proposed optimized BitTorrent-like contend replication framework, we performed a performance evaluation campaign, by using an ad-hoc

[7]We recall that they are typical heuristic policies implemented by many common p2p file-sharing systems that aims at improving the robustness of the swarm, as presented in Section 5.5..

In particular, each peer selects the chunks to be uploaded (i.e., the vector $h_i(t)$) starting from the rarest among all the peers. The peers to which the chunks are uploaded (i.e., the vector $r_i(t)$) are random as in (i).

software simulator. Then, we investigated two scenarios, and we introduced congestion in the network in order to study if there are any impacts in the overall system performance when the system relies on bandwidths that are not strictly reflecting the real available capacity. In the following, we will use FAIR-RND and FAIR-RF to denote the "fair" algorithms with random and rarest-first strategy for the choice of the network, respectively. Besides, with OP2P we denote the proposed optimized heuristic. For performing simulations with the optimized system we set the parameter $L = 20$. Concerning the "fair" strategies we present results obtained by computing the average over 100 runs corresponding to different feasible networks. For all the simulations we assumed the scenarios are "steady", i.e., churn is absent. Readers interested in performance of the system when in presence of the churn could find more interesting details in [77].

Specifically, in Scenario 1 we adopted a mixed environment, composed by two classes of peers. The first class is composed by 6 peers (including the *seeder*) equipped with a slow ADSL connectivity (i.e., 512 kbit/s for the downstream and 256 kbit/s for the upstream) and 4 peers with a high speed symmetrical access of 4 Mbit/s both for upstream and downstream traffic. Conversely, in Scenario 2, we introduced a mixed environment, as presented in Table 8.1.

Table 8.1. Access characteristics of peers participating in the swarm of scenario 2.

Peer Number	Downstream	Upstream
1 *(Seeder)* - 6	512 Kbit/s	512 Kbit/s
7 - 15	1 Mbit/s	1 Mbit/s
16 - 20	512 Kbit/s	256 Kbit/s

Regarding the congestion, we assumed that each connection (i.e., at the transport level) has a probability p at each stage t of being routed by the network through a congested path. In our simulations we chose p in such a way that no more than 30% of the NK total chunks are sent via a congested connection at any time t. The transfer loss parameter $\gamma_{i,s}$ for any peer i and any congested connection s has been taken equal to 0.8, which is representative of a congested network that reduce users' capacities without completely stalling their transfers. Besides, owing to the presence of the TCP as a transport protocol, we modeled the congestion as a speed reduction, without introducing any data loss phenomena.

Figure 8.2 depicts the trends of the cost functions grouped for each scenario. The cost functions for each "fair" strategy represent a given run selected from the pool of 100 runs. It can be noticed that the cost functions for the OP2P strategy decrease quicker than the others, resulting in shorter time for seeding a content over the entire architecture. Figure 8.2 also shows that the OP2P achieves better gains when compared with the FAIR-RF. In fact, the FAIR-RF triggers peers in the swarm. Thus, remote peers tend to synchronize by requesting concurrently the rarest chunk, hence quickly saturating the rarest chunk owner and preventing from distributing the load among the distribution swarm. In addition, Figures 8.2(a) and 8.2(b) highlight that the OP2P strategy outperforms the "fair" ones even in presence of congestion issues.

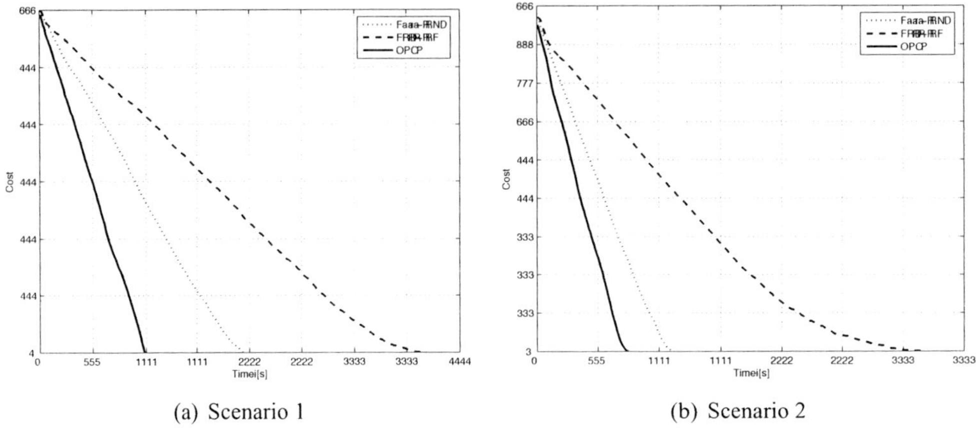

(a) Scenario 1 (b) Scenario 2

Figure 8.2. Cost trends of different scenarios.

8.2.3. Remarks on the Optimization Applied to File-Sharing

As it has been presented in Section 8.2.1. and Section 8.2.2., optimization of p2p systems is a valuable tool for the purposes of rationalizing resources and enhancing the perceived quality of service, without changing the underlying network characteristics (e.g., the bandwidth provisioning).

As shown, in modern tracker-based content delivery platforms (e.g., BitTorrent), the tracker could be used to compute organized strategies to optimize some metrics (e.g., bandwidth usage, policies of the replication process, ...) of the distribution swarm.

Besides, another important application of optimization aims at controlling queues in file-sharing systems, in order to enhance the system fairness (as it has been explained for an eMule-like system). Therefore, by using an optimization framework and applying optimal control strategies, such as predictive control, it is possible to enhance the overall system fairness.

However, the peculiarities of the p2p file-sharing environment rises some problem when optimizing [78].

Specifically:

- p2p systems are affected by churn. This implies that the state of the system changes during the time. Then, there is the need of sampling it in order to optimize over a fixed-sized state. In fact, by considering the overlay flavor of p2p systems, jointly with the need of orchestrating the interactions, reflect in mixed-integer problems. Such problems are computational intensive. Besides, the two aforementioned objectives are difficult to achieve simultaneously: in fact, short sample times, imply high performance algorithm to complete the computation before the Δt expires;

- when in presence of real optimizations, the cost could have piecewise flat trends and the space could be vast. In this perspective, tools like Monte-Carlo methods are interesting and valuable to solve such large-scale problems;

- analyzing other control mechanisms, such as those proposed in [79], [80] and [81], is a needed step to gain more comprehension on the control techniques best suited for p2p file-sharing applications.

Chapter 9

Behavior Analysis of File-Sharing Systems

Behavior analysis of p2p file-sharing applications is a mandatory step for many actors. For instance, network administrators struggle with traffic patterns to understand how to tame the resulting load. Besides, it is also important to avoid the endangering of traffic engineering policies employed to maintain a satisfactory level of QoS. In addition, academics are developing measurement methodologies and metrics to better quantify impact of such applications over the network and to produce synthetic models to conduct simulative investigations. As a first step, we present a quick survey about other traffic studies available in the literature with the related major findings. Also, we propose the basic tools employed to gather comprehension of traffic produced by p2p file-sharing applications. Then, interesting metrics are given as well.

The core of the chapter summarizes a behavior and traffic analysis performed over an eMule community, available in [68]. We decided to present this analysis since, being focused on well-defined internals, it could be adopted as a reference to tweak and to engineer such applications.

9.1. A Quick Survey of the Literature

As partially introduced in Chapter 1, the reason of the impressive adoption of p2p technologies is rooted within two major causes: the Internet has reached many homes, and broadband accesses offered via flat-rate contracts allow users to dare with ever bigger contents [82]. Consequently, the impact of file-sharing over the Internet is non negligible, and such application has generated a relevant attention in the literature.

As described in [3], which is a fundamental work about the traffic analysis of file-sharing services over a large scale system, file-sharing originated traffic is an increasing fraction of the overall Internet traffic. In addition, [83] and [84] present analyses of traffic gathered in edge routers, offering a detailed sketch, even if local, of p2p usage.

Since new services become suddenly available, many studies focused on systems no longer existing or now abandoned in favor of the "cool system of the day". The reasons for declining popularity may depend on the specific systems. The most recurrent motivations are: the system has been forced to shutdown, users discovered cycle stealing software within the client [85], or more effective software became available. In this respect, [9], [86] and [87] focused on the Gnutella system [88], which is no longer a mainstream, even if further developments and analyses are ongoing [89]. The literature extensively reports different studies conducted on p2p file-sharing related traffic, also focusing on a wide range of network scenarios. In [90] the status of the increasing inter-domain traffic among ISP networks is analyzed and correlated with the presence of file-sharing systems. Furthermore, [91] presents a comprehensive traffic analysis of the Greek School Network (GSN), registering that 50% of the outgoing and 37% of the incoming traffic is generated by file-sharing applications.

As underlined in [3], many analyses are solely focused on the signaling traffic of file-sharing systems, neglecting the traffic patterns generated by file transfer operations. An interesting analysis of the signaling traffic, both via analytical tools and simulations, is proposed in [92]. Concerning the modeling of file-sharing systems, and their evaluation via simulation tools, a thorough analysis can be found in [93]. Besides, an interesting investigation, even if not totally based on file-sharing, is available in [94]; it extensively explains the impact of fragmented traffic on the Internet.

Regarding existing systems, two major file-sharing frameworks are today widely utilized: BitTorrent and eMule. BitTorrent has gained popularity since it has been extensively adopted not only as a mere file-sharing program, but also as an effective content delivery system, being able to handle *flashcrowds*, which, we recall, are sudden peaks of simultaneous requests [62]. BitTorrent's traffic has been analyzed, and detailed studies can be found in [95] and [96].

Moreover, the eDonkey 2000 [97] file-sharing system, which, as presented, is the direct ancestor of the eMule system, has been investigated. Concerning traffic analyses specifically performed on this file-sharing framework, some preliminary work is available [98]. As to emphasize its timeliness, its portability over standard General Packet Radio System (GPRS) [99] or Universal Mobile Telecommunications System (UMTS) [100] has been evaluated.

Lastly, many of the works focusing on the collection of data at the user's side, rather than being able to infer users' behaviors by observing traffic, clashed on objective difficulties, such as those due to the need of crawling and gathering data in a highly mutable large scale system. As a matter of fact, the majority of user-side analyses have been performed on BitTorrent, owing to the presence of a centralized component, called the tracker [63], which allows to collect many statistics with a very simple framework.

9.2. Tools and Techniques for File-Sharing Analysis

In this section we show the two main approaches adopted to investigate the behavior of a file-sharing application. Specifically, we will introduce *active* and *passive* methods. Besides, we complete the portrait by presenting a novel index to study and classify traffic produced by p2p file-sharing applications and some techniques presented in many client interfaces preventing a straightforward traffic collection and analysis.

To avoid overlaps, in Section 9.3. the analysis of a Nation-wide eMule community will be performed, by using a technique (i.e., the adoption of an instrumented client interface jointly with a centralized data repository) that is not covered in this section.

9.2.1. Active Methods

Concerning active methods, we mention *crawlers*. They have been originally introduced to perform an automatic browse of the WWW. In this case, they are also known as spiders, as to emphasize their relationship with the "web". Such software entities reside on a machine, and they are used to fetch many information about the web, e.g., how pages are connected through hyperlinks. The gathered metrics are commonly used to gain some knowledge of the web, in order to organize and exploit it to perform some operations. The most relevant scope, is to use crawler to index web pages, allowing search engines to work.

Crawlers are also employed to investigate many aspects of an overlay used by p2p file-sharing applications [101]. For instance, the snapshots in Figure 2.5 located in Section 2.3.3. have been created by using a crawler. As regards file-sharing applications, they are often employed to collect topology information.

Employing a crawler for collecting information about a p2p-created overlay poses the following challenges:

- the number of nodes could be huge, as well as the amount of information to be stored;
- owing to the presence of churn, the gained snapshot is not always accurate.

Both issues are also present when developing spiders. However, the severe values of churn highly account for a reduction of accuracy. To face such issues, crawlers are also developed by relying upon parallel and distributed systems able to perform several operations concurrently. In fact, by reducing the time employed to crawl the overlay, there are chances that the number of topological changes that happen due to churn are reduced.

To develop a crawler of a p2p file-sharing application, the following requirements have to be met:

- having a list of peers to feed the crawling infrastructure, or at least to kickstart the crawl procedure;
- implementing the protocol (or the needed subset of functionalities) in order to contact remote peers and gain the needed information;
- storing the data in a rational manner for subsequent analyses.

For instance, the crawler employed in [9] was aimed at acquiring a snapshot of the Gnutella overlay. To do so, developers implemented the discovery mechanisms based on the Ping-Pong (as described in Section 3.3.1.) to collect topology information. In a nutshell, a crawler is then an automated client-interface being able to contact remote nodes and to exchange data with them in order to explore the network.

Notice, that a crawler could be used to collect a plethora of information[1], allowing to investigate different aspects of a p2p file-sharing application. For instance, it could used,

[1]Obviously, both the protocol specification and the client interface implementation must be able to collect and deliver the needed information request. For instance, if the protocol does not support requesting the available bandwidth, other mechanisms must be employed to collect such value.

upon implementing the needed portion of the protocol, to ask the number of shared files, the type of connection, the number of incomplete downloads, the uptime, ...

9.2.2. Passive Methods

As regards passive methods, we present here *sniffers*, that is the short name for packet or network sniffers. They are employed to capture data streams over a network. For instance, they can be used on a machine to collect the outbound and inbound traffic, or by carefully inserting probes within a network (e.g., by replicating traffic by using a switch and delivering it over a sniffing machine) to gain understanding of the flows circulating in the network[2]. In addition, there are also sophisticated agent-based system that could be placed in routers to collect data and statistics in geographic and world-wide deployments.

Sniffers are the main tool to collect and analyze network traffic. Thus, they are also fundamental to investigate file-sharing applications. For instance, to gain comprehension of specific applications, to reverse engineer the internals, or to check the implementation of a brand new application, a developer usually runs the client interface on a target machine and starts to analyze the produced traffic patterns.

One of the most successful network sniffers, is Wireshark[3]. Gaining confidence with this software is a good starting point to obtain comprehension of traffic generated by file-sharing applications. In fact, it has pre-built filters, being able to isolate flows produced by the most popular systems and to parse the produced PDUs, thus recognizing searches, pings, transfers and so on.

We point out that sniffers are not solely employed to collect data for off-line analyses. Rather, they are also employed to perform on-line monitoring, e.g., to feed Intrusion Detection Systems (IDS).

9.2.3. The Content Transfer Index

Collecting and analyzing the traffic produced by a p2p file-sharing service is not substantially different from doing the same for other network applications. However, due to the high volumes and the heterogeneous characteristics of the involved client interfaces, this task could be more complicated.

In fact, file-sharing applications usually produce two different kinds of traffic: *signaling* (e.g., to perform searches and to maintain the overlay, as explained in Chapter 2 and to exploit core operations as explained in Chapter 3) and *data transfers* (e.g., to actually move the parts composing the file among remote peers).

---

[2]Actually, deciding where to place probing machines and what to collect is a non-trivial task. In fact, it heavily depends on how the network is engineered. Placing a standard commodity machine devoted to sniffing could be impractical in high speed link, since the internal buses (such as the PCI) could constitute a bottleneck. In this perspective, the sniffing machine would drop packets. Conversely, in IEEE 802.3 based LANs, their broadcast nature could allow to sniff packets of the entire network from a standard machine placed almost everywhere in the network, i.e., without the need of acting over the configuration of the network. However, to collect the traffic generated by a large-scale distributed system, such as a p2p file-sharing application, commodity hardware is often unsuitable, and more sophisticated systems have to be employed.

[3]`http://www.wireshark.org/`. Notice that this program is formerly known as Ethereal.

Thus, it is of interest being able to quantify the contribution to the overall traffic load of this two different components. Of course, it is possible to program a protocol analyzer or to rely on preexistent solutions, for instance those presented in [102].

In this vein, Bolla, Canini, Rapuzzi and Sciuto, in their work aimed at characterizing the traffic of p2p file-sharing services, which is available in [103], noticed that such approach suffers from the following major drawbacks: i) to be exploited, a deep understanding of the protocol internals of the specific p2p file-sharing application to be investigated is needed; ii) payloads of each packet have to be evaluated and iii) the "state" of every flow encountered has to be maintained, thus impeding the scalability of the approach.

Then, they propose a novel index, called Content Transfer Index (CTI) in order to discriminate between traffic belonging to signaling or data transfers. Such index has been engineered to classify a couple of data flows[4], i.e., the superposition of data flowing from an host to another and viceversa, among two endpoints.

The $CTI \in [0,1]$, of a conversation C is defined as follows:

$$CTI = \frac{F}{f+F} \cdot \frac{\bar{P}}{MSS(C)} + \frac{f}{f+F} \cdot \frac{\bar{p}}{MSS(C)}$$

where, F and f are the lengths of the two flows composing the conversation C, with $f \leq F$, $\bar{P}$ and $\bar{p}$ are the average number of bytes of the payload of packets composing the flows with length F and f, respectively. $MSS(C)$ is the Maximum Segment Size (MSS) of the conversation C if TCP is employed. Besides, if the data flow is embodied within UDP, $MSS(C)$ becomes:

$$MSS(C) = MTU(C) - dim(IP_{header}) - dim(UDP_{header})$$

where, $MTU(C)$ is the Maximum Transfer Unit observed in the UDP flow composing the conversation, and $dim(IP_{header})$ and $dim(UDP_{header})$ are the header lengths of the IP and UDP packets, respectively[5].

The CTI is then simple, yet effective. The intuition under its development is that when peers receive packets filled up to the MTU and send a smaller number of packets, typically this is a data transfer and the "reverse" flow is due to acknowledgments employed by the TCP. Besides, signaling flows consist of "balanced" conversations, i.e., they have a flatter profile.

Authors experimented the CTI with datasets of eMule traffic and they found that imposing a CTI threshold equal to 0.2 allows to correctly classify the 95% of download conversations.

9.2.4. Techniques Preventing a Seamless Traffic Analysis

Besides the reasons previously explained (i.e., scalability and volumes), traffic analysis of p2p file-sharing applications could be even more complicated due to particular techniques

[4]Actually, Authors do prefer to call the data flow a "conversation". But, the concept of "conversation", which is discussed in [103], is very similar to endpoints defined in Section 4.2.1..

[5]Notice that this formalism is not present in the original work available in [103], but it is introduced here for didactical purposes.

implemented in the client interfaces. In fact, as a direct consequence of their "borderline" nature[6] in the sense of legality, p2p file-sharing client interfaces implemented several countermeasures to prevent blocking and dissimulating the produced traffic patterns.

Specifically, techniques preventing a seamless collection of the traffic generated by file-sharing applications and, consequently an effective analysis are:

- random port hopping: this is a simple but popular technique. After a TCP connection has been established, the data flow is periodically moved from the current TCP port to another. In addition, when a peer attempts to connect to the network, it first tries a TCP port and if the connection attempt fails, it tries another one, randomly selected from a prefixed port pool. Obviously, knowing the port range utilized allows this approach to be compensated, but sometimes a large set of ports is adopted. In addition, sometimes well-known ports are used: consequently, the traffic generated by file exchanges could be a not relevant part of the traffic observed over a given port, thus reflecting in unneeded volume collected. Besides, peers, in order to be able to successfully "hop" from a port to another, have to: i) use a fixed pattern or ii) use some kind of signaling to announce the next TCP port to be used. In this perspective, if i) is employed, studying the "hop sequence" is enough to avoid analyzing not needed traffic. If ii) is employed, it is possible to build some adaptive monitoring tools being able to observe/collect conversations. Needles to say, this approach could rise some technical challenges;

- data encryption: probably, this is one of the most popular countermeasures adopted within many file-sharing applications. It is based on peers ciphering the traffic flow, in order to prevent pattern recognition; such technique has been also introduced not only to augment the level of privacy of users, but also to enrich the security properties of the overall file-sharing service. For instance, data encryption has been used to avoid traffic spoofing to force peers to leave the overlay or to have a misbehaving attitude;

- HTTP tunneling: it is a technique usually adopted when a firewall impedes traffic flow. An HTTP tunnel offers the possibility of encapsulating data inside an HTTP flow and route it through a firewall. Usually, web browsing is allowed in organization networks and for home users: consequently, the firewall is configured to let the HTTP traffic flow. This technique brings to an abuse of port 80, but it allows masquerading not allowed traffic under the guise of normal web browsing. Naturally, the amount of traffic reveals the true nature of the communication. But it forces an observer to deal with web traffic that could have relevant volumes and many conversations, thus accounting for scalability issues during the phase devoted to gain comprehension. In addition, as explained in Section 8.1.1., embodying data within HTTP could also be used to take advantage of ISPs' webcaches deployed for standard Web-browsing. In this perspective, the overall observed volume could not exactly reflect the real phenomena of file exchange activities. We point out that HTTP or HTTP-like protocols are commonly employed also as signaling methods in file-sharing applications, for

[6]As hinted to in Chapter 1, file-sharing networks usually convoy files reflecting in many copyright infringements.

instance to perform handshaking with other peers or supernodes, search queries or download requests.

9.3. Analysis of a Nation-Wide eMule Community

In Section 8.1.2. we presented the modifications performed to the standard eMule client interface to support peculiarities of a nation-wide user community. By cooperating with mod developers, we were able to analyze the traffic volumes produced by users in the aforementioned setting. In this perspective, we report here the measurement methodology and the most important results.

We believe that discussing this topic is important for the process of engineering file-sharing applications, since it allows to understand how to monitor a file-sharing application in order to: i) adjust its design, thus performing major engineering and ii) understand the internals when deployed over a large-scale scenario. Interested readers will find more details and results in [68].

9.3.1. Measurement Methodology

The measurement methodology adopted, relied on an instrumented client interface, which communicates behaviors of interest (that will be presented in detail in Section 9.3.3.) to a centralized repository[7]. The reasons under this choice are:

- while developing and engineering a client-interface, it is fundamental to understand the performance and the effectiveness of the different subsystems. Such information could be absent in the produced traffic patterns, or difficult to find out while observing aggregate behaviors;
- file-sharing developers could not be able to "reach" the network, thus collecting traffic patterns is impeded. In addition, the wide diffusion of the service might account for an overlay spanning over several ISPs. Therefore, having the chance of collecting data from several ISPs simultaneously (maybe world-wide and with different policies) could be unfeasible.

Then, the client interface has been coded to gather the statistics of interests and to deliver the data to a centralized repository. The latter has been coded as a simple server-side application, i.e., listening to remote connections from client interfaces and then store results over a DB. This "collection facility" has been connected to the Internet via a standard user access (specifically, a FTTH access at 10 Mbit/s). As regards the DB employed to store and process data, a SQL-like [104] database and ad-hoc scripts in Python [105] have been utilized. Such software set-up has been running over commodity hardware (i.e., a standard PC).

[7]Notice that this method differs from those presented in Section 9.1.. In fact, presented works were almost based on sophisticated analysis methodologies and relied on mining a very remarkable amount of data. Yet, about the totality of the presented works relied solely on the observation of traffic flows collected over the network.

Notice that this set-up has very low requirements in terms of hardware (it runs on commodity hardware), software (it relies on open source technologies) and connectivity (it uses a standard SOHO Internet access). In this perspective, by exploiting the nature of p2p ("action" takes place in end-nodes) it is possible to understand important behaviors without accessing the underlying network infrastructure. This is also why "user oriented" activities aimed at understanding and developing p2p file-sharing applications can take place, thus boosting the effectiveness of file-sharing services.

Then, a special version of the eMule mod has been delivered to end users. Such version was modified to collect end user data and send them to the aforementioned centralized system for the storage. The delivery of the software was quite simple, since the modified eMule client interface adopted by this community has a feature for on-line version checking and automatic update. Since a user can override the updater and postpone indefinitely the download and the installation of the new version, eMule has been further modified in order to reject connections from too old clients, as to encourage user to utilize up-to-date clients[8]. Nevertheless, the option responsible of activating the collection of data has to be explicitly selected. The collected dataset utilized in this work was provided by the 99% of the overall community[9].

9.3.2. Privacy Issues

When gathering data, some challenges in the field of the privacy are always present. For instance, when gathering traffic directly from the network, IP addresses potentially reveal the identity of the parties involved in the conversation. In our case, being the owners of the "centralized repository" not directly involved in the maintenance of underlying network infrastructure (i.e., they are users rather than network administrators), such aspect is mitigated. Anyway, having data collected and bind with their IP addresses normally reduces the degree of participation.

But, as explained in Section 6.7. users and resources are identifiable within the eMule overlay. Then, to ensure a proper degree of privacy[10], data has been collected by relying upon a mechanism allowing to uniquely identify an eMule client interface, without being able to recognize it.

In fact, at the first run, the modified eMule generates another hash, $HASH_{stat}$, differing from the $HASH_{user}$, which is utilized to identify the peer within the eMule network. Moreover, $HASH_{stat}$ is only utilized by the eMule client and the server collecting the statistics; hence it is not propagated within the overall network. In this manner, it is impossible to perform "reverse" look-up, i.e., recovering the peer identity from the stored data[11].

[8]Actually, also the standard eMule client interface has a built-in function to reject connections from peers running an outdated version of the software. However, here we refer to an additional check performed on the modified eMule software under investigation.

[9]For more details about the user participation, please refer to [68].

[10]We believe that users will cooperate if enough privacy is provided. Even if we are not able to quantify this statement, this is the perception of people constantly moving around the development of file-sharing applications.

[11]Of course, this happens if the developers of the centralized repository "play fair". In fact, if you are experienced with socket programming, as soon as an incoming `connection request` is received, upon performing an `accept()` it is possible to seamlessly retrieve the remote IP address.

9.3.3. Collected Data

Roughly, we processed data from millions of user sessions, and each session contains several performance indicators, e.g., information on the exchanged traffic volumes, how some preferences have been set by users, and other miscellaneous statistics.

In order to handle the amount of data, without resorting to sophisticated DBs, cumulative values have been stored. Then, the subsequent analysis will rely on cumulative values. However, this coarse-grained method allows to investigate different aspects: i) understanding the knowledge that could be gained by independent developers without the cooperation of network administrators; ii) proving that it is possible to analyze some file-sharing behaviors even relying on "poor" data (i.e., not directly observed from network traffic); iii) showing that some findings are consistent with other analyses performed with more complicated techniques, for instance those proposed in Section 9.1..

The data collected from eMule clients, and how they are organized within the database will be briefly described.

- $HASH_{stat}$: as previously explained, a brand new hash, used to identify the peer that produces the session data.
- *Connection Type:* the type of connection adopted by the user. Three values are possible: Fiber, ADSL, or Unknown.
- *Share Ratio Enforcer:* identifies whether the enforcer defined in Section 8.1.2. is active or not.
- *Cumulative Total Upload*: the total traffic uploaded by the peer.
- *Cumulative Total Download*: the total traffic downloaded by the peer.
- *Cumulative Upload within the Community*: the total traffic uploaded to community members.
- *Cumulative Download within the Community*: the total traffic downloaded from within Community members.
- *Upload Value*: the value set by the user for the maximum upload.
- *Upload Allocation Percentage*: how the upload bandwidth has been allocated among external/community users when the enforcer defined in Section 8.1.2. is overruled.
- *Version*: the version of the eMule client producing the statistics.
- *Run Time*: how long the eMule peer remained connected.
- *Update Times*: time-stamps, indicating when the specific client interface updated the session data.
- *City*: the city where the peer is located.

9.3.4. A Remark on IP Addressing Inspection

Concerning the key collecting the peer's city, it has been introduced in order to find out if some spatial correlation is present, i.e., trying to understand if exchanged volumes follow some particular distribution over different MANs. There are several techniques to find out the peer's location in a precise manner, for instance by using geo-location services [106] or ad-hoc procedures [107]. However, we were able to retrieve the peer location in a simple and effective manner. Actually, the ISP maps the city into the first octet of the private IP space. Owing to the ISP architecture, locating a MAN is equivalent to locating the city[12]. The ISP also utilizes a special IP numbering, in order to identify a user accessing the Internet via FTTH or DSL. For this reason, we were able to identify the access technology for about the totality of the eMule clients. However, a very small subset of clients were unable to retrieve their IP address and then identify their location within the ISP. The main issues preventing a client interface to retrieve its address is due to the presence of multiple NATs and firewalls; in a nutshell, it is due to the lack of transparence. Notice also, that this fact prevents, or at least rises the complexity of developing "location aware" p2p file-sharing applications, thus requiring some proper countermeasures.

This is well representative of how the knowledge available in peers populating the file-sharing network (i.e., their ability of recognizing and delivering their IP addresses) could be exploited also to find out some internals of the underlying network infrastructure. Even if the book will not present the analysis of the spatial distribution of the traffic volumes within the ISP, it has been proposed to showcase an important engineering option. In fact, if it is possible to retrieve information about the numbering scheme employed by the ISP[13], notice how it can be exploited to empower the overlay topology matching discussed in Section 2.2.1..

9.3.5. Effectiveness of the eMule Credit System

Figure 9.1 depicts the scatterplot of all the eMule peers and their overall traffic volumes produced. The plot is in log-log scale, and it can be useful to analyze the inference of the credit systems (e.g., like those described in Section 5.2.1. and Section 5.2.2.) or of an incentive framework, such as those of [46], [108] and [109].

Specifically, the plot has been produced by organizing all the eMule peers in the $(\log(volume_{up}), \log(volume_{down}))$ space and also depicting the 1:1 ratio boundary, which is representative of peers having $(\frac{Up}{Down})_{Traffic} = 1$.

The Figure clearly shows that the presence of a credit system is noticeable, since peers' traffic tends to be attracted by the straight line. However, we can observe that for low volumes (e.g., $\leq 10^{-2} Gigabytes$), the credit system has difficulties in attracting peers, resulting in a scattered plot. For relevant data volumes, the peers appear to be polarized

[12]This is why the terms city and MAN are used interchangeably throughout this chapter.

[13]There are several methods to achieve that. It is possible to do a kind of "reverse engineering", for instance by means of an instrumented client interface reporting both the IP address and the content of some "form" filled by users. Notice in this case that the "correctness" and willingness to cooperate of users play a role, as well as the presence of NATs in the network, as discussed in Chapter 4. Besides, the ISP numbering scheme could be documented or available to you, for instance because you are in charge of developing a file-sharing application for the ISP itself.

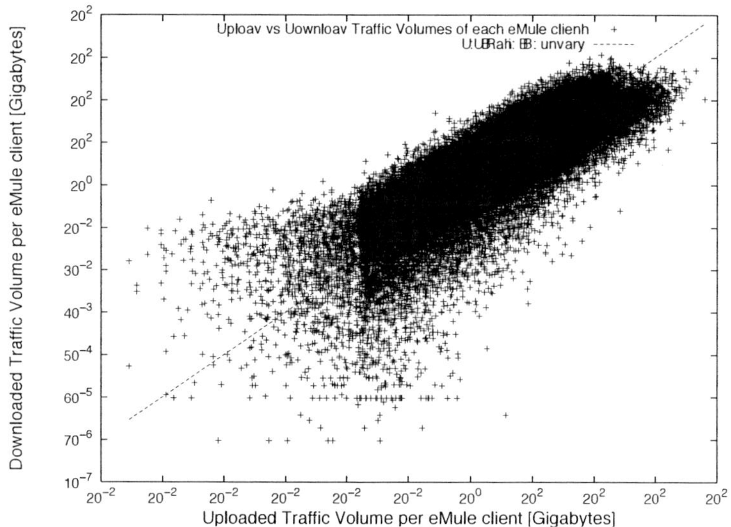

Figure 9.1. Scatterplot of all the eMule peers participating in the community: uploaded vs downloaded volumes. Axes are in log-log scale. Also depicted the 1:1 ratio boundary, that separates eMule peers with a share ratio < 1 and > 1, respectively.

across the 1:1 boundary. We point out that many peers are still in the zone characterized by $(\frac{Up}{Down})_{Traffic} < 1$ and also here the same considerations made for the previous analysis apply.

9.3.6. Consideration on the Effectiveness of Enforcers

Figure 9.2(a) depicts the Cumulative Distribution Function (CDF) of various upload values set by DSL eMule users. The CDF curve evidences that DSL users tend to assign a relevant amount of the upstream bandwidth to eMule, highlighting a correct perception of the access system. In fact, the upstream bandwidth, being completely decoupled from the downstream one, cannot be utilized to download contents; hence, users tend to maximize its exploitation as a "means of exchange".

It is also important to underline that values $\leq$ 10 kbyte/s are almost never set. This is due to users wanting to avoid to have the download bandwidth policed, according to mechanism presented in Section 5.3. and Section 6.5..

Figure 9.2(b) depicts the CDF of various upload values set by FTTH eMule users. The curve highlights a barely adequate bandwidth contribution of FTTH eMule users. There is a sharp change in its shape around the value $\sim$ 850 kbyte/s, revealing the presence of mainly two user classes. One class is representative of fair users, while the other one is representative of very impressive contributors.

Also for FTTH users, upload values $\leq$ 10 kbyte/s are almost never set, as previously explained. Almost all eMule users (both DSL and FTTH) appear to be conscious of the bandwidth enforcement introduced by eMule and they want to maximize the utilization of their download bandwidth, by setting the upload bandwidth accordingly. Therefore, an overall consideration can be made: file-sharing applications engineers and mod developers

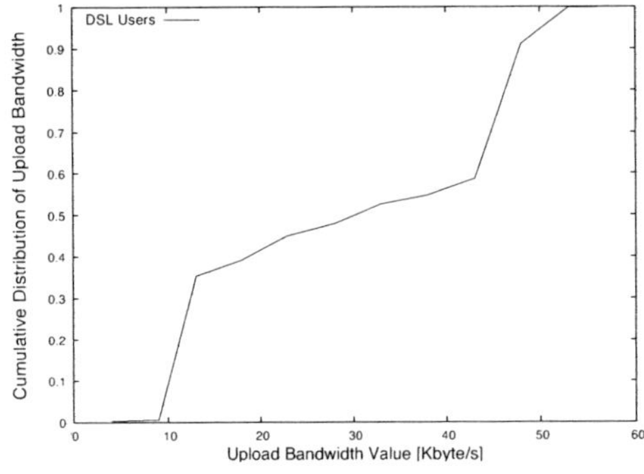

(a) Cumulative distribution of upload bandwidths of DSL users. The upload values have been partitioned in 5 kbyte/s steps.

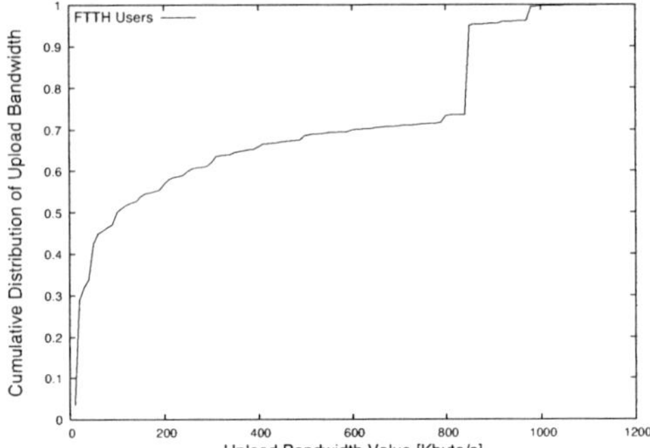

(b) Cumulative distribution of upload bandwidths of FTTH users. The upload values have been partitioned in 10 kbyte/s steps.

Figure 9.2. Cumulative distributions of upload bandwidths of eMule users.

wanting to push users to share the bandwidth could exploit this fact, which appears to be "convincing" for the users.

9.3.7. On The Usage of Zipf's Law

The most widely adopted traffic and content distributions in the literature related to p2p and file-sharing are based on the Power-Law [16], [110] and the Zipf 's Law [111]. Power-Law is strictly related to the Zipf's distribution, being its generalized formulation. In Section 2.3.2. we discussed how the Power-Law has been used to model the distribution of the overlay connectivity in unstructured p2p networks.

Zipf's Law allows to analyze data according to the *rank-frequency* criteria. Put briefly, rank-frequency distributions allow to study the occurrence frequency of objects within a fixed set. Such distributions have been introduced for lexical analysis and subsequently adopted for other purposes.

The Zipf's Law states that if f_r is the frequency of a word of a fixed vocabulary (say Ω), then $f_r \propto \frac{1}{r}$, where r is the rank, i.e., f_1 is the most frequent word, f_2 is the second one, and so on. In a bi-log scale, its representation is given by a line with a slope equal to -1. The Power-Law distribution generalizes such concept, and defines a distribution having $f_r \propto \frac{1}{r^{\psi}}$. In a bi-log scale, its shape will be a line with slope equal to $-\psi$. However, many studies have risen doubts on the validity of such law to describe the data distribution of all p2p file-sharing applications. At the best of our knowledge, the first study showing that p2p traffic does not obey any power law rule is [112], subsequently supported by [3]. Particularly, [112] highlighted this behavior for the Kazaa [15] system, while [3] highlighted it also for Kazaa, Gnutella and Direct Connect [44]. The reported works were based on traffic collected over the network and not in end-nodes and, moreover, they focused on several file-sharing systems, but not on the eMule one.

Also in our analysis, we investigated such distributions in three different cases: overall traffic volumes created by eMule users over DSL, over FTTH and, finally, the aggregated scenario (i.e., without making distinctions between DSL and FTTH access technologies).

The results are similar to that of [3], since the distributions of traffic generated within sessions by eMule peers are not straight lines in the log-log scale.

Thus, our analysis, that is solely based on data collected directly from client interface, supports similar results of previous work relying upon more "sophisticated" data collection strategies.

Concluding, the adoption of the Zipf's Law to model different aspects of modern p2p file-sharing applications should be carefully evaluated, depending on the particular context.

Chapter 10

Dissecting and Modifying Applications

Starting from Chapter 1 we introduced a quick historical portrait of file-sharing applications. Then, through Chapters 2 to 9 we entered in the algorithmic and technical details of p2p file-sharing software.

In this chapter we "close the loop" by inspecting some parts of the source code at the basis of the most popular file-sharing client interfaces. This effort is twofold: on one hand it allows to recognize where algorithms discussed in this book are placed within the overall software architecture, as well as how they are implemented. On the other hand, it helps to understand where to introduce modifications, tweaks and hooks to collect data, thus performing research, measurements and to modify some basic behaviors.

Lastly, investigating source code written by other developers is a valuable effort to start writing a new file-sharing application.

10.1. Why to Dissect and Modify Applications

Studying the source code of a preexistent application always gives precious hints on how-to implement new services from scratch, or to understand behaviors of interests.

Specifically, in the case of file-sharing applications this accounts for:

- recovering to the lack of technical documentation;
- understanding if particular aspects of traffic patterns are triggered by some well-defined algorithms, coding choices, erratic implementations or misbehaviors in the ported code;
- modifying applications to log particular events, packets and control variables (e.g., the values of modifiers discussed in Chapter 5) to study the performance of the system, to collect data in real settings (e.g., to feed synthetic models for simulation purposes) or, generally speaking, to conduct research;

- tweaking the subsystems presented in Chapter 3 to enhance their response or to apply results obtained through optimization (for instance, to implement the code needed to exploit mechanisms described in Chapter 8);
- enhancing the underlying protocols by adding extensions;
- developing new features to adapt the client interface to correctly behave in different network settings. For instance, to deploy a preexistent file-sharing service also in unforeseen network architectures (see for instance the solutions proposed in Section 8.1.2.) or to introduce algorithms to overcome the lack of transparency (as discussed in Chapter 4).

10.2. Analysis of the eMule Source Code

We start the discussion about the analysis of source code of file-sharing client interfaces by firstly inspecting the eMule system. At the time of writing, the last stable version of eMule is v.0.49a, therefore we show snippets of such release. The source code for the Windows operating system is freely available[1] in the download section of the official eMule web site (`http://www.emule-project.net/`).

10.2.1. Management of the Upload Queue

Let us analyze the code responsible of implementing the logic devoted to the management of the upload queue. A said, eMule's upload queue is ruled through a credit system (explained in Section 6.6.). The core of the code is located in the files `UploadQueue.cpp` and `UploadQueue.h`.

There are many methods[2] belonging to the class `CUploadQueue`, which is responsible to perform the needed operations on queued clients. For instance, there are methods to parse the list populated by data structure describing the queued peers, find top ranking clients, add upload slots (if needed), update the status of the Graphical User Interface (GUI) and so on. We showcase here the most interesting and didactical one as to foster the understanding of a p2p file-sharing client interface internals.

The first method we showcase is

```
CUpDownClient* CUploadQueue::FindBestClientInQueue();
```

which is devoted to find the highest ranking peer in the waiting queue. Such client is then returned via a pointer to `CUpDownClient`, defined in `BaseClient.cpp`. The method also contains the logic to deal with LowID-ed clients. In fact, a LowID-ed peer that is not connected, but would be ranked highest if connected, is identified via a proper flag, which means that the client should get an upload slot as soon as it connects.

[1] As it happens for client interfaces running on other operating systems.

[2] The original eMule client interface is coded in C++. Thus, *methods* is the correct nomenclature. However, since we are proposing a "high level" analysis of the sources, we will sometimes interchange the words functions, methods, procedures to keep the discussion enough clear also to Object Oriented Programming (OOP) newcomers.

Also, let us introduce a method employed for managing peers in the first K positions of the queue, i.e., managing the upload slots (see Section 6.6.1. for more details about the upload slots).

Specifically

```
void CUploadQueue::Process();
```

adds and removes queued peers from the upload queue. Besides, it is also responsible to interact with the *socket architecture* of the eMule client interface. Such layer has been introduced to abstract the different flavor of communications (e.g., server oriented, UDP-based, webservice, ...) employed by the system. Interested readers will find the implementation details in files available in the source tree containing the word `Socket`, e.g., `UDPSocket.cpp` and `UDPSocket.h`. Thus, the `Process();` method also guarantees that the socket architecture has the transmission queues always filled with a proper amount of data (i.e., packets).

Another important operation performed in this portion of code is the one devoted to calculating the amount of bytes sent, in order to effectively exploit *bandwidth throttling* for each served peer. In fact, as explained in Section 6.6., the available upload bandwidth has to be shared (fairly) among the K available slots.

To do that, two functions are employed:

```
GetNumberOfSentBytesSinceLastCallAndReset();

GetNumberOfSentBytesOverheadSinceLastCallAndReset();
```

which allow to quantify both the data and the overhead bytes stuffed into each TCP connection belonging to a slot, respectively. Notice that, owing to the presence of a layered architecture to access the needed TCP sockets, the latter have been properly encapsulated. In this vein, such functions are two excellent examples of where to put some hooks to log data allowing to gather traffic information. Moreover, notice that it is then possible to collect such "high level" information (i.e., the number of sent bytes) without resorting to intercepting data in the path towards the sockets.

The `Process();` method is called approximatively each 100 ms and, consequently, also the measures available via the previous functions are clocked with a "tick" with the same granularity. By inspecting the method implementation, we notice that the software only stores, for each peer owning a slot, the last 30 seconds history, as well as a portion of code devoted to quantify the amount of data uploaded to friends via the friend slot (as explained in Section 6.6.1.).

Lastly, we point out the following method is also responsible of adding a client interface to the remote queue. Also, it preliminary makes also some checks, such as: i) check if this client[3] interface is LowID-ed; ii) check if the peer is connected through the Kad network in order to trigger the proper callback mechanism; iii) check if the peer is a friend to adjust as soon as possible its position in the queue and spawn a friend slot if needed; iv) check if

[3] In the sense of the client interface running the portion of code, not a remote one.

this client interface is connected to the same server of the remote one, as to inform other subsystem of the possibility of performing a TCP callback to connect LowID-ed peers (as explained in Section 6.4. and in Chapter 4) and v) check the length of the queue to avoid bandwidth trashing.

The function is:

```
void CUploadQueue::AddClientToQueue(CUpDownClient* client,
bool bIgnoreTimelimit);
```

We point out the boolean parameter `bIgnoreTimelimit`, which accounts for checking/ignoring if the requestor exhibits a too aggressive behavior in terms of number of requests in a given timeframe, thus resulting in a ban.

The next code snippet highlights the risk of IO bottlenecks on the disk by having too many synchronized activities at once. As said, the code managing the queue is executed 10 times in a second. Periodically, the data collected from queued peers (i.e., the list of clients, the file storing credits, ...) has to be saved in order to guarantee the persistence of the information through sessions.

```
// one second
if (counter >= 10)
{
counter=0;
theApp.clientcredits->Process(); // 13 minutes
theApp.serverlist->Process(); // 17 minutes
theApp.knownfiles->Process(); // 11 minutes
theApp.friendlist->Process(); // 19 minutes
theApp.clientlist->Process();
theApp.sharedfiles->Process();
}
```

In this perspective, the data is synced to disk and is evaluated once every second, as to avoid the waste of CPU time (see the `if` statement in the previous snippet). However, many concurrent updates of the files storing needed information (e.g., the `credits.met` file for the credits, the `server.met` file for the servers and so on) could account for bottlenecks. Thus, besides managing in a wise way the different threads composing the client interface, it is also important to reduce concurrencies in IO by de-sincying reads and writes. To achieve that, eMule developers save different files in different times.

Also, one might argue that, with modern hardware, the concurrent write of small files could endanger the performances of the IO subsystem. However, even if some buffering mechanism is already implemented within the client interface, remember that sending/receiving different files simultaneously already stresses the disk subsystem.

Lastly, besides the properties granted by using an OOP language, we highlight that having a well-defined separation of subsystems (as explained in Chapter 3) allows to easily extend and modify the application. This is the case of the dual-queue architecture introduced in the modified eMule version discussed in Section 8.1.2..

10.2.2. Bandwidth Throttling

In Section 8.2.2. we discussed how to optimize BitTorrent-like applications. A fundamental requirement was to be able to allocate a given bandwidth share to a specific connection. Even if we consider the eMule client interface, rather than the BitTorrent one, we can investigate mechanisms exploited here to throttle the bandwidth. Besides, as deeply discussed in Section 6.5., eMule also relies on enforcers and caps of the available bandwidth. Nevertheless, the needed fair sharing of bandwidth across the upload slots accounts for the need of some “rate limiting” scheme.

Thus, having an effective mechanism to throttle the upload bandwidth is mandatory for this kind of applications and for their further enhancements. As a matter of facts, eMule has an effective throttling system, which is implemented in the files `UploadBandwidthThrottler.cpp` and `UploadBandwidthThrottler.h`.

Basically, the *upload throttler* runs in a separate thread and it receives data from the filesystem feeding it to the socket architecture in a limited and “shaped” way. In a nutshell, it is an high level scheduler responsible of shaping the rate accordingly to the alloted bandwidth for a specific connection/data flow. The technique employed is to stuff a proper amount of bytes in a given time frame to reflect in a “regularized” data rate of x kbyte/s. Such a mechanism assures a quite precise behavior if observed in an adequate time frame, since it is subject to local glitches. Also notice that to achieve a proper degree of precision, this method should rely on sufficient small time frames, thus reflecting in a thread that has to be put in execution very often. This also accounts for CPU usage, which could also be significant for a relevant amount of upload connections that have to be throttled. The constructor is `UploadBandwidthThrottler::UploadBandwidthThrottler(void);` and it responsible of spawning the thread running the throttling code.

The constructor is depicted in the following snippet:

```
UploadBandwidthThrottler::UploadBandwidthThrottler(void) {
m_SentBytesSinceLastCall = 0;
m_SentBytesSinceLastCallOverhead = 0;
m_highestNumberOfFullyActivatedSlots = 0;

threadEndedEvent = new CEvent(0, 1);
pauseEvent = new CEvent(TRUE, TRUE);

doRun = true;
AfxBeginThread(RunProc, (LPVOID)this);
}
```

Notice a set of variables, which quantifies how many bytes have been sent between two subsequent executions of the throttler function (e.g., `m_SentBytesSinceLastCall`) and a statement to create and manage the thread. In addition, the class has also the proper accessor methods such as:

```
uint64 GetNumberOfSentBytesOverheadSinceLastCallAndReset();
```

and those for adding/removing socket to be throttled. Specifically, we showcase the following function:

```
void UploadBandwidthThrottler::AddToStandardList(uint32 index,
ThrottledFileSocket* socket);
```

Such method is responsible of adding sockets to be throttled that already own an upload slot. Sockets are sorted in a list, which is processed sequentially from the first socket (having and index equal to 0) and so on.

The two parameters are used to have some control over the process of inserting the socket in the list. Specifically, the 32 bit integer `index` states the index where to place a given socket. If the value of `index` is greater than the number of active sockets, then it means that such socket has to be put last in the list. Besides, the parameter `socket`, which is a pointer to a `ThrottledFileSocket`, indicates the socket to be added to the list in order to be properly managed.

We underline the relevant presence (and need) of an efficient threading architecture, since the client interface must exploit many duties concurrently. Therefore, threading is important to maintain the code clear, even if application could be more difficult to debug. As a remark, we invite the reader to analyze implementation of different file-sharing applications written in C/C++ or in Java, and the different support and implementation of their threading architecture.

10.2.3. The Kad Network

As discussed in Section 6.3., eMule also relies on a DHT-based structured overlay called Kademlia, or simply Kad. The implementation source is somewhat "split" from the rest of the sources, and it is placed in the subdirectory `kademlia`.

The Kad source tree is organized as follows:

- `io`: it contains functionalities for managing IO operations devoted to read/write files, and access to buffered operations;
- `kademlia`: this portion contains logic to handle and perform searches, to interact with the eMule GUI and to test if the UDP traffic has been firewalled;
- `net`: it realizes all the functionalities needed to receive and keep track of the UDP traffic. Thus, it is also responsible of parsing the data contained in the PDU delivered via the UDP protocol: for instance, `KADEMLIA_PUBLISH_REQ`;
- `routing`: it exploits the routing strategy (e.g., calculating the XOR distance and managing bins) at the basis of the Kademlia algorithm. In addition, it uses basic filters (e.g., guaranteeing only one KadID per IP) allowing to protect the network from attacks or misbehaviors;

- `utils`: it implements utility functions to enhance and support basic operations, such as Big Endian to Little Endian conversion, IP addresses to string translation and so on.

Having the Kad implementation well-separated from the rest of the application is handy to make local changes without the need of adjusting the rest of the code. This is why it has been possible to tweak[4] the Kad implementation for particular network settings without the need of redesigning the rest of the application. For instance, as it has been done for the modified eMule presented in Section 8.1.2.. Besides, it seems also a needed step, since Kad has been added later in the life-span of the eMule client interface.

It is also worth to notice that Kad is basically employed to search content and nodes, thus its main role is to exploit operations described in Section 3.5.1.. Then, it can rely on the other functionalities already available in the client interface (e.g., file handling).

Moreover, we point out that also the "Kad layer" spawns its own threads, that have to be properly "synchronized" with the rest of the application.

10.2.4. Low Level Code

We briefly report here a remark on low-level optimization that could be useful to do some CPU intensive operations. In addition, this allows to highlight the benefits offered by the work of the open source community.

As explained in Section 6.7., eMule, likes many other client interfaces, uses both MD5 and SHA functions to identify resources. Computing such hashes could be time intensive, for instance if a huge amount of files are shared or if the dimension of the file is relevant (and sometimes the two conditions are met simultaneously). Therefore, eMule relies on optimized code written in x86 assembly to compute such values. The implementation is available in `MD4_asm.asm` and `SHA_asm.asm`. Such code has also to be used with a proper .cpp "interface", available in the files `MD4.cpp`, `MD4.h`, and `SHA.cpp`, `SHA.h`, respectively.

We point out that both implementations, owing to the source availability, have been borrowed from the Shareaza file-sharing client interface, which is available at the URL: `http://www.shareaza.com/`.

10.3. Analysis of the BitTorrent Source Code

To complete our investigation of source code analysis of most popular file-sharing applications, we briefly investigate specific functionalities of the original BitTorrent implementation, written in Python. Such client interface is very popular, even if also the Azureus one is becoming on of the most widely adopted[5]. Source code of the BitTorrent client interface written in Python are available at the URL `http://www.bittorrent.com/`.

[4]From the comments in the Kademlia source code: *"Changing something without knowing what all it does can cause great harm to the network if released in mass form"*. In fact, modifying the behavior of highly cooperative applications could bring the entire overlay in danger, so it has to be carefully evaluated.

[5]While writing, the Azureus client interface has been enriched with a multimedia interface to act both as a sort of multimedia content provider and player. Besides, this evolution has been enriched with search capabilities, to prevent the need of browsing the web or ad-hoc repository to find out .torrent files. Even if Azureus still is well kwon as a BitTorrent client interface, it has been renamed as Vuze, as to emphasize its new soul, i.e., not solely being a p2p file-sharing application.

As explained in Chapter 7, such system relies on two different components: the tracker and the client interface. Since the tracker is a simple centralized entity, basically exploiting a webservice, we will omit the analysis of its source code here. Notice that modern programming environments such as the aforementioned Python have a rich set of frameworks to implement server-side and web-based applications in a simple and straight manner. Therefore, this is the reason why we focus our attention solely on the client interface.

10.3.1. NAT Handling

As explained in Chapter 4, it is important to detect and react when in presence of NAT devices, as to avoid misbehaviors in the overall p2p architecture. To do that, proper functionalities have been implemented in two modules, called `NatCheck.py` and `NatTraversal.py`, which are responsible of detecting and traversing NAT devices, respectively. It is interesting to note the use of UPnP (to have a brief insight about UPnP, please see Section 4.4.) to exploit traversal. Particularly, the module is responsible of creating a thread and to handle requests/responses while performing the proper signaling and configurations. We underline the portion of code related to the Simple Object Access Protocol (SOAP), which creates ad-hoc templates to handle UPnP communications[6].

For instance, consider the following snippet taken from `NatTraversal.py`:

```
get_mapping_template = ('<?xml version=''1.0"?>' +
'<s:Envelope xmlns:s="http://schemas.xmlsoap.org/soap/envelope/"' +
's:encodingStyle="http://schemas.xmlsoap.org/soap/encoding/">' +
'<s:Body>' +
'<u:GetGenericPortMappingEntry xmlns:u=' +
'"urn:schemas-upnp-org:service:WANIPConnection:1">' +
'<NewPortMappingIndex>%d</NewPortMappingIndex>' +
'</u:GetGenericPortMappingEntry>' +
'</s:Body>' +
'</s:Envelope>')
```

Therefore, the proposed modules are candidate to be extended or tweaked to provide traversal support to client interfaces. Nevertheless, to conduct experiments or gather data with NAT deployed in the "wild", logging data and events here could be a wise starting point.

Then, to avoid misunderstanding we do prefer to focus our analysis on a "pure" BitTorrent client interface. Interested readers are encouraged also to check Vuze's source code since, differing from eMule, it is coded in Java. As to the other client interfaces investigated in this chapter, the source code is freely available in the download section of the official Azureus website (`http://azureus.sourceforge.net/`) and, at the time of writing, the latest stable version is v.3.1.1.0.

As a final remark, its source tree is well organized and separated into logical "groups". The biggest two concern the GUI and the specific logic implementing the actual p2p file-sharing operations, which is called "the core". Within the core, there is an another level of subdivision, i.e., functions for logging, IP addresses filtering, peer management, NAT traversal, download operations, and so on.

[6] Also fun is the remark in the code saying "`if you think for one second that I'm going to implement SOAP in any fashion, you're crazy`". Actually, there are more elegant ways to exploit SOAP, but this works efficiently.

We also point out the importance of relying over standards (e.g., the HTTP protocol for convey communications) and having a coding environment that supports in a simple and efficient way the creation/parsing of URLs and HTTP-based communications, such as the Python one.

10.3.2. Rate Limiter

Being able to limit the rate of TCP flows employed for exchanging files is mandatory for the correct exploitation of techniques presented in Chapter 5. As already explained for the eMule bandwidth throttling (see Section 10.2.2.), also BitTorrent has similar mechanisms. Even if coded in a different language, the rough technique is the same, i.e., data to be sent through sockets is scheduled via a well tamed thread. Therefore, to avoid overlaps, we will omit redundant technicalities. However, we give hints about how to exploit optimized and tracker-computed strategies to optimize the overall p2p service, as explained in Section 8.2.2..

Recall that in the proposed optimized framework, the tracker computes strategies both in terms of *defining the peers to connect to* and the *bandwidth shares* to assign to a given TCP connection (which relies on the peer wire protocol, as presented in Section 7.4.).

In order to actually exploit such a framework, it is mandatory to:

- quantify the effective amount of bandwidth, in order to: i) provide a feedback to the scheduler within the client interface to properly adjust allocations and ii) send a feedback to the decision maker, i.e., to avoid bandwidth trashing;
- alter dynamically the bandwidth allotted for a given TCP flow, i.e., "programming" on the fly the scheduler with the directive computed by the tracker. Obviously, a proper signaling mechanism must be in place, but extending the BitTorrent tracker protocol via extensions (presented in Section 7.3. and in a more detail here [65]), or by creating ad-hoc solution over HTTP, is quite straightforward.

To accomplish with the previous requirements, it is possible to start the investigation of the two modules `RateMeasure.py` and `RateLimiter.py`, which measure and limit the rate of a data flow to be delivered through the network, respectively.

The `RateMeasure.py` module has several methods to quantify the stuffed data, and the code is quite simple. Besides the `RateLimiter.py` specifies that data is sent to peers by employing a round-robin discipline.

Let us consider the following snippet (which has been cleaned from remarks):

```
def set_parameters(self, rate, unitsize):
if unitsize > 17000:
unitsize = 17000
self.upload_rate = rate
self.unitsize = unitsize
self.lasttime = bttime()
self.offset_amount = 0
```

notice the possibility of setting the `rate`, which could be parsed from a remote communication with the tracker, as proposed in Section 8.2.2..

Another important detail is given by the `if` statement limiting the `unitsize` quantity to 17000; notice that such a variable describes the data unit employed by the BitTorrent peers to exchange data with each other (as presented in Section 7.4.). Such limit has to be introduced due to the round-robin nature of the scheduler enforcing the limits. In fact, the scheduler is designed to serve at most one full request per round. Increasing such limit would reflect in sending more data to peers issuing request with a size larger than the 16 Kbit. Thus, forcing the limit to 17.000 reflects in enough bandwidth to have metadata to pass without excessive delays.

Chapter 11

Conclusions

As said, p2p file-sharing applications allow to exchange a huge amount of data in a cooperative and scalable way, supporting up to millions of users. In this chapter we present a final summary about the covered topics, and also we briefly propose some additional activities (called here "exercises") to experiment and gain more comprehension of file-sharing applications. Lastly, interested readers are highly encouraged to dig into the proposed bibliography or, at least, to read such works that have been referred as "seminal" through the book.

11.1. A Final Summary

In Chapter 1 we quickly analyzed the history of file-sharing applications and their evolution towards p2p architectures, as a method to avoid lawsuits and to augment their efficiencies. Besides, we showcased the most successful p2p file-sharing services.

Then, in Chapter 2 we introduced the basic notions to gain comprehension of the p2p communication paradigm, highlighting both its weaknesses and strengths. In addition, we investigated the different flavors of overlay networks employed for developing such services, and their major properties and overheads.

By using the previous ones as a foundation, in Chapter 3 we thoroughly discussed the functional and architectural components needed to deploy a fully featured file-sharing application. We discussed about the different "customizations" needed to maintain the file-sharing services operative, and we also underlined critical operations, such as those to kick-start a file-sharing infrastructure. In this vein, it is clear that some centralized or well-defined components are needed, thus somewhat reducing the extremely distributed flavor of p2p file-sharing applications.

In Chapter 4 we investigated the issues that arise when deploying file-sharing applications over real network deployments, i.e., the Internet. Therefore, we discussed the most popular traversal techniques and methodologies used to recover the lack of transparency affecting the modern Internet (or the near future one, when IPv6 and IPv4 will remain in place simultaneously).

Owing to the highly autonomic nature of users, (hence, peers populating the overlay), in Chapter 5 we analyzed the most effective and deployed techniques to support cooperation in file-sharing applications.

From this point, we started what could be defined, at least virtually, the *second* part of the book. In fact, we moved from general (but not unrealistic) concepts to two real-world case-studies. Specifically, Chapter 6 and 7 proposed the analysis of two of the most popular file-sharing applications, eMule and BitTorrent, respectively.

Then, we moved into the *third* part, and we analyzed two different efforts aimed at modifying or enhancing file-sharing systems. In fact, in Chapter 8, modifications to the eMule system developed by users populating the "underground" of the Internet, have been discussed. In particular, we focused our attention on modifications to support a nation wide p2p community. Moreover, we discussed the need of applying optimization techniques to file-sharing services to augment the overall degree of performance, and to balance (if not to tame) specific behaviors, such as the degree of fairness of credit systems. Such research topic has been also discussed to give the reader some examples about how to model some mechanisms acting as the basis of modern file-sharing client interfaces.

Another important topic, which shows the impact of file-sharing applications over the network, has been introduced in Chapter 9. Here, we proposed the behavior analysis of the eMule file-sharing application, highlighting some interesting behaviors (e.g., the effectiveness of the credit system) and, thus motivating and explaining concepts introduced earlier in the book. Besides, an overall view about traffic analysis, in terms of advancements performed in the literature and interesting solutions and tools have been proposed.

As a natural conclusion of the dissertation, and as a last tool needed to be able to autonomously follow a path to do major engineering of p2p file-sharing applications, we proposed the analysis of the implementations of the most relevant client interfaces and some details on how to code and modify some specific protocols.

This concludes the preliminary effort to climb the learning curve for engineering file-sharing applications. We started from theory and we ended into implementation details. We want to recall that there are many algorithms and ideas available in the "library" of the scientific community that are waiting to be deployed into file-sharing systems to reach the next level.

Now, it is up to you to continue the journey in to this world.

Good luck with file-sharing.

11.2. Exercises

1. Try to identify the kind of overlay employed in file-sharing applications you know.

2. After installing a sniffer, are you able to understand the traffic generated by the client interface?

3. Choose an open source file sharing application. Are you able to find out the portion of the code implementing the protocol?

4. Choose a closed source file sharing application. Are you able to reverse engineer the protocol? Are you able to understand how peers are contacted without accessing the protocol specification?

5. Try to selectively block traffic on specific ports, or specific protocols (e.g., UDP and TCP), then analyze the behavior of the client interface. Can you understand the effect of limiting the end-to-end connectivity on the client interface?

6. Analyze the behavior of a file-sharing application with and without a NAT in the middle. Did you notice differences? Con you discover any mechanism to bypass NAT devices by inspecting traffic patterns?

7. Many client interfaces allow logging. Can you understand the output of the log of the most popular file sharing applications?

8. Choose an open source application. Are you able to build a very simple network application able to connect to a file-sharing network?

9. Use and keep running a file-sharing application (i.e., eMule) for a while. Then, close it and start to observe the traffic. Are there some traffic patterns even if your client interface is not running anymore? Why?

10. Are you able to understand and map the different components of available file-sharing applications according to concepts introduced in Chapter 3?

References

[1] W. T. Hollyday (Ed.), "Governmental Principles and Statutes on Child Pornography", Nova Science Publishers, 2003, ISBN 1-59033-846-4.

[2] V. G. Cerf, "Musing on the Internet", EDUCAUSE, September - October 2002, pp. 74 - 84.

[3] S. Sen, J. Wang, "Analysing Peer-To-Peer Traffic Across Large Networks", *IEEE/ACM Transactions on Networking,* April 2004, Vol. 12, No. 2, pp. 219 - 232.

[4] T. Karagiannis, A. Broido, N. Brownlee, K. C. Claffy, M. Faloutsos, "Is P2P Dying or Just Hiding?", in *Proc. of IEEE Globecom 2004 - Global Internet and Next Generation Networks,* Dallas, TX, November 2004, Vol. 3, pp. 1532 - 1538.

[5] S. Bradner, "Omniscience Protocol Requirements", *IETF RFC* **3751**, April 2004, available on-line at: `http://www.ietf.org/rfc/rfc3751.txt?number=3751`.

[6] "JXTA v2.0 Protocols Specification", available on-line at: `http://www.jxta.org`.

[7] E. Rosen, A. Viswanathan, R. Callon, "Multiprotocol Label Switching Architecture", *IETF, RFC* **3031**, January 2002, available on-line at: `http://www.ietf.org/rfc/rfc3031.txt?number=3031`.

[8] T. Oh-ishi, K. Sakai, K. Kikuma, A. Kurokawa, "Study of the Relationship between Peer-to-Peer Systems and IP Multicast", *IEEE Communications Magazine*, Vol. 41, pp. 80 - 84, January 2003.

[9] M. Ripeanu, I. Foster, "Mapping the Gnutella Network", *IEEE Internet Computing,* Vol. 6, pp. 50 - 57, January 2002.

[10] Y. Chu, S. Rao, S. Seshan, and H. Zhang, "Enabling Conferencing Applications on the Internet using an Overlay Multicast Architecture", in *Proceedings of ACM SIGCOMM,* August 2001, pp. 55 - 67.

[11] Y. Chu, S. Rao, and H. Zhang, "A Case for End System Multicast", in *Proceedings of ACM SIGMETRICS,* June 2000, pp. 1 - 12.

[12] S. Fahmy, M. Kwon, "Characterizing Overlay Multicast Networks and Their Costs", *IEEE/ACM Transactions on Networking*, Vol. 15, No. 2, pp. 373 - 386, April 2007.

[13] D. Stutzbach, R. Rejaie, "Understanding Churn in Peer-to-Peer Networks", in *Proceedings of the 6th ACM SIGCOMM Conference on Internet Measurement (IMC 2006)*, Pisa, Italy, September 2006, pp. 189 - 202

[14] S. Ghemawat, H. Gobioff, S.-T. Leung, "The Google File System", *ACM SIGOPS Operating Systems Review*, Vol. 37, No. 5, pp. 29 - 43, 2003.

[15] J. Liang, R. Kumar, K. W. Ross, "The KaZaA Overlay: A Measurement Study", in *Proceedings of the 19th IEEE Annual Computer Communications Workshop,* 2004, Bonita Springs, Florida, October 2004.

[16] L. A. Adamic, R.M. Lukose, A.R. Puniyani, B.A. Huberman, "Search in Power Law Networks", *Physical Review E,* Vol. 64, pp. 046135-1 - 046135-8.

[17] S. Saroiu, P. Krishna Gummadi, S. D. Gribble, "A Measurement Study of Peer-to-Peer File Sharing Systems", in P*roceedings of Multimedia Computing and Networking (MMCN'02)*, San Jose, CA, January 2002.

[18] Q. Deng, H. Lv, "Analyzing Unstructured Peer-to-Peer Search Networks with QIL", in *Proceedings of IEEE International Conference in Services Computing (SCC'04)*, September 2004, Shanghai, China, pp. 547 - 550.

[19] Napster Protocol Specification, available on-line at: `http://opennap.sourceforge.net/napster.txt`.

[20] The OpenNap - Open Source Napster Server, official Homepage: `http://opennap.sourceforge.net/`.

[21] I. Stoica, R. Morris, D. Karger, M. Kaashoek , H. Balakrishnan, "Chord: A Scalable Peer-to-Peer Lookup Service for Internet", in *Proceedings of the Conference on Applications, Technologies, Architectures, and Protocols for Computer Communications*, San Diego, CA, August 2001, pp. 149 -160.

[22] D. Karger, E. Lehman, M. Levine, D. Lewin, R. Panigrahy, "Consistent Hashing and Random Trees: Distributed Caching Protocols for Relieving hot Spots on the World Wide Web", in *Proceedings of the 29th Annual ACM Symposium on Theory of Computing*, El Paso, TX, May 1997, pp. 654 - 663.

[23] D. Lewin, "Consistent Hashing and Random Trees: Algorithms for Caching in Distributed Networks", Master's Thesis, Department of EECS, MIT, 1998. Available at the MIT Library, and also on-line at: `http://thesis.mit.edu/`.

[24] L. Caviglione, F. Davoli, "An Overlay Scheme to Provide Loose QoS for Wireless Nodes", in *Proceedings of the International Workshop on Broadband Wireless Access for ubiquitous Networking (BWAN'06)*, Alghero, Italy, September 2006.

[25] L. Caviglione, F. Davoli, M. Polizzi, S. Bellisario, "Handling Local User Mobility and QoS in a Controlled Ad-Hoc Environment", in *Proceedings of the Australasian Telecommunication Networks and Application Conference (ATNAC 2007),* Christchurch, New Zealand, December 2007.

[26] A. Williams, "Requirements for Automatic Configuration of IP Hosts", IETF Internet Draft, draft-ietf-zeroconf-reqts-12.txt, September 2002, available on-line at: http://tools.ietf.org/html/draft-ietf-zeroconf-reqts-12.

[27] L. Caviglione, L. Veltri, "Using SIP as a P2P Technology", *African Journal of Communication and Information Technology*, Vol. 1, No. 1, September 2005, pp. 38 - 43.

[28] L. Caviglione, L. Veltri, "Using SIP as P2P Technology", *Invited Paper at 12th International Conference on Telecommunications (ICT 2005)*, Capetown, South Africa, May 2005.

[29] L. Caviglione, "The Dark Side and the Force of the Peer-To-Peer Computing Saga", p2p Journal, pp. 1-11, January 2004.

[30] R. Rivest, "The MD4 Message-Digest Algorithm", *Network Working Group, RFC* **1320**, IETF, April 1992, available on-line at: http://www.ietf.org/rfc/rfc1320.txt?number=1320.

[31] V. Fuller, T. Li, J. Yu, K. Varadhan, "Classless Inter-Domain Routing (CIDR): an Address Assignment and Aggregation Strategy", *Network Working Group, RFC* **1519**, IETF, September 2003, available on-line at: http://www.ietf.org/rfc/rfc1519.txt?number=1519.

[32] B. Ford, P. Srisuresh, D. Kegel, "Peer-to-Peer Communications Across Network Address Translators", in *Proceedings of the of the annual conference on USENIX Annual Technical Conference*, Anaheim, CA, USA, pp. 13 - 18, 2005.

[33] S. Guha, P. Francis, "Characterization and Measurement of TCP Traversal through NATs and Firewalls", in *Proceedings of Internet Measurement Conference (IMC)*, Berkeley, CA, pp. 199 - 211, October 2005.

[34] P. Srisuresh, M. Holdrege, "IP Network Address Translator (NAT) Terminology and Considerations", *Network Working Group, RFC* **2663**, IETF, August 1999, available on-line at: http://www.ietf.org/rfc/rfc2663.txt?number=2663.

[35] K. Egevang, P. Francis, "The IP Network Address Translation", *Network Working Group, RFC* **1631**, IETF, May 1994, available on-line at: http://www.ietf.org/rfc/rfc1631.txt?number=1631.

[36] The NatTrav Software Library, Version 1.00 [Alpha], available on-line at: http://sourceforge.net/project/shownotes.php?release_id=613159.

[37] R. Stevens, "UNIX Network Programming", Vol. 1, Second Edition: Networking APIs: Sockets and XTI, Prentice Hall, 1998, ISBN 0-13-490012-X.

[38] J. Liang, R. Kumar, K.W. Ross, "The FasTrack Overlay: A Measurement Study", *Computer Networks (Special Issue on Overlay Distribution Structure and their Applications)*, Vol. 50, No. 6, April 2006, pp. 842 - 858.

[39] The Universal Plug and Play Forum, Homepage: http://www.upnp.org/.

[40] Internet Gateway Device (IGD) Standardized Device Control Protocol V 1.0, available on-line at: http://www.upnp.org/standardizeddcps/igd.asp.

[41] J. Rosenberg, R. Mahy, P. Matthews, "Traversal Using Relays around NAT (TURN): Relay Extensions to Session Traversal Utilities for NAT (STUN)", Network Working Group, draft-ietf-behave-turn-09, IETF, July 2008, available on-line at: http://tools.ietf.org/html/draft-ietf-behave-turn-09.

[42] J. Rosenberg, J. Weinberger, C. Huitema, R. Mahy, "STUN - Simple Traversal of User Datagram Protocol (UDP) Through Network Address Translators (NATs)", Network Working Group, RFC 3489, IETF, March 2003, available on-line at: http://www.ietf.org/rfc/rfc3489.txt?number=3489.

[43] L. Caviglione, F. Davoli, "Satellite Communications and Peer-to-Peer Networking: Architectural Hazards and Interoperability Issues", in *Proceedings of the 10th Ka and Broadband Communications Conference*, Vicenza, Italy, October 2004, pp. 285 - 292.

[44] Direct Connect (DC++), Homepage: http://dcplusplus.sourceforge.net/.

[45] M. L. Garcia Osma, F. J. Ramon Salguero, G. Garcia de Blas, J. Andres Colas, J. Enriques Gabeiras, S. Perez Sanches, R. Trueba Fernandez, "Enabling Local Preference in Peer-to-Peer Traffic", *COST* **279** TD(04)017, Technical Document, 2004.

[46] B. Cohen, "Incentives Build Robustness in BitTorrent", available on-line at: http://bitconjurer.org/BitTorrent/bittorrentecon.pdf.

[47] G. Neglia, G. Lo Presti, H. Zhang, D. Towsley, "A Network Formation Game Approach to Study BitTorrent Tit-for-Tat", in *Proceedings of EuroFGI International Conference on Network Control and Optimization*, Avignon, France, June 2007.

[48] M. Flood, "The Princeton Mathematic Community in the 1930s", Transcript Number 11, Trustees of Princeton University, 1985.

[49] R. Axelrod, "The Evolution of Cooperation", New York: Basic Books, ISBN 0-465-02121-2.

[50] R. Axelrod, "The Evolution of Cooperation Revised Edition", Perseus Books Group, ISBN 0-465-00564-0.

[51] Wikipedia page with the formulation of the Prisoner's Dilemma, available on-line: http://en.wikipedia.org/wiki/Prisoner's_dilemma.

[52] Limewire, Homepage: http://www.limewire.com/.

[53] L. Garces-Erice, E. W. Biersack, P. A. Felber, K. W. Ross, G. Urvoy-keller, "Hierarchical Peer-to-Peer Systems", in *Proceedings of ACM/IFIP International Conference on Parallel and Distributed Computing (Euro-Par)*, Rennes, France, 2003, pp. 643 - 657.

[54] G. Tan, S. A. Jarvis, X. Chen, D. P. Spooner, G. R. Nudd, "Performance Analysis and Improvement of Overlay Construction for Peer-to-Peer Live Media Streaming", in *Proceedings of the 13th IEEE International Symposium on Modeling, Analysis, and Simulation of Computer and Telecommunication Systems (MASCOTS)*, Atlanta, Georgia, USA, 2005, pp. 169-178.

[55] M. Bishop, S. Rao, K. Sripanidkulchai, "Considering Priority in Overlay Multicast Protocols Under Heterogeneous Environments", in *Proceedings of the 25th IEEE International Conference on Computer Communications (INFOCOM 2006),* Barcelona, Spain, April 2006, pp. 1 - 13.

[56] S. Serbu, S. Bianchi, P. Kropf, P. Felber, "Dynamic Load Sharing in Peer-to-Peer Systems: When some Peers are more Equal than Others", *IEEE Internet Computing Special Issue on Resource Allocation*, Vol. 11, No. 4, July/August 2007, pp. 53-61.

[57] D. Carra, E. W. Biersack, "Building a Reliable P2P System Out of Unreliable P2P Clients: The Case of KAD", in *Proceedings of ACM CoNext*, New York, NY, USA, December 2006.

[58] Y. Kulbak, D. Bickson, "The eMule Protocol Speficiation", *Leibniz Center Technical Report* TR-2005-03, School of Computer Science and Engineering, The Hebrew University, 2005.

[59] P. Maymounkov, D. Mazieres, "Kademlia: a Peer-to-Peer Information System Based on the XOR Metric", in *Proceedings of the 1st International Workshop of Peer-to-Peer Systems*, Cambridge, MA, March 2002.

[60] Official Kademlia, Homepage: `http://kademlia.scs.cs.nyu.edu/`.

[61] D. Eastlake, P. Jones, "US Secure Hash Algorithm 1 (SHA1)", IETF, RFC 3174, September 2001, , available on-line at: `http://www.ietf.org/rfc/rfc3174.txt?number=3174`.

[62] A. Bharambe, C. Herley, V. N. Padmanabhan, "Analyzing and Improving BitTorrent Performance", *Technical Report* MSR-TR-2005-03, Microsoft Research, February 2005.

[63] M. Izal, G. Urvoy-Keller, E. Biersack, P. Felber, A. Al-Hamra, L. Garces-Erice, "Dissecting BitTorrent: Five Months in a Torrent's Lifetime", in *Proceedings of the 5th International Workshop on Passive and Active Network Measurement (PAM 2004)*, Antibes Juan-les-Pins, France, April 2004.

[64] Wikipedia page about Bencode: `http://en.wikipedia.org/wiki/Bencode`.

[65] The BitTorrent Protocol Specification V.1.0, available on-line: `http://wiki.theory.org/BitTorrentSpecification`.

[66] The Onion Router Project Homepage: `http://www.torproject.org/`.

[67] X. Yang, G. de Veciana, "Service capacity of peer to peer networks", in *Proceedings of the 23rd Annual Joint Conference of the IEEE Computer and Communications Societies (INFOCOM 2004)*, March 2004, Hong Kong, Vol. 4, pp. 2242 - 2252.

[68] L. Caviglione, F. Davoli, "Traffic Volume Analysis of a Nation-Wide eMule Community", *Computer Communications Journal,* Elsevier, Vol. 30, No. 10, June 2008, pp. 2485 - 2495.

[69] L. Caviglione, C. Cervellera, F. Davoli, F. A. Grassia, "Optimization of an eMule-like Modifier Strategy", *Computer Communication Journal*, Elsevier, `doi:10.1016/j.comcom.2008.04.005`.

[70] L. Caviglione, C. Cervellera, "Design of a Peer-to-Peer System for Optimized Content Replication", *Computer Communications Journal*, Elsevier, Vol. 30, No. 16, November 2007, pp. 3107-3116.

[71] D. Q. Mayne, J.B. Rawlings, C.V. Rao, P.O.M. Scokaert, "Constrained Model Predictive Control: Stability and Optimality", *Automatica* **36**, 2000, pp. 789 - 814.

[72] G. S. Fishman, "Monte Carlo: Concepts, Algorithms, and Applications", Springer, 1996.

[73] L. Caviglione, C. Cervellera, F. Davoli, F. A. Grassia, "An eMule-like Modifier Strategy Based on Optimization", in *Proceedings of the International Symposium on Performance Evaluation of Computer and Telecommunication Systems (SPECTS 2007)*, San Diego, CA, USA, July 2007.

[74] H. Niederreiter, "Random Number Generation and Quasi-Monte Carlo Methods", SIAM, PA, 1992.

[75] L. Caviglione, C. Cervellera, "Design and Performance Evaluation of an Optimized Peer-to-Peer Content Replication Scheme for Vehicular Networks", in *Proceedings of the International Symposium on Performance Evaluation of Computer and Telecommunication Systems (SPECTS 2008)*, Edinburgh, UK, June 2008.

[76] R. Morris, "TCP Behavior with many Flows", in *Proceedings of the 1997 International Conference on Network Protocols (ICNP 1997)*, Atlanta, GA, 1997.

[77] L. Caviglione, C. Cervellera, "A Peer-to-Peer System for Optimized Content Replication", *European Journal on Operation Research*, Elsevier, `doi:10.1016/j.ejor.2008.03.048`.

[78] L. Caviglione, C. Cervellera, "On the Optimization of Peer-to-Peer Systems", in *Proceedings of the 41st Hawaii International Conference on System Sciences (HICSS-41)*, Kona, Hawaii, USA, January 2008.

[79] G. Bartolini, E. Punta, T. Zolezzi, "Approximability Properties for Second-Order Sliding Mode Control Systems", *IEEE Transactions on Automatic Control*, Vol. 52, No. 10, pp. 1813 - 1825, 2007.

[80] L. Levaggi, E. Punta, "Analysis of a Second Order Sliding mode Algorithm in Presence of Input Delays", *IEEE Transactions on Automatic Control*, Vol. 51, No. 8, pp. 1325 - 1332, 2006.

[81] G. Bartolini, E. Punta, T. Zolezzi, "Simplex Methods for Nonlinear Uncertain Sliding mode Control", *IEEE Transactions on Automatic Control,* Vol. 49, No. 6, pp. 922 - 933, 2004.

[82] N. Christin, A. S. Weigend, J. Chuang, "Content Availability, Pollution and Poisoning in File Sharing Peer-to-Peer Networks", In *Proceedings of the* 6th *ACM Conference on Electronic Commerce,* Vancouver, Canada, June 2005, ACM Press, New York, NY, pp. 68 - 77.

[83] N. Leibowitz, A. Bergman, R. Ben-Shaul, A. Shavit, "Are File Swapping Networks Cacheable? Characterizing P2P Traffic", in *Proceedings of the 7th International Workshop on Web Content Caching and Distribution*, Boulder, CO, August 2002.

[84] S. Saroiu, K. P. Gummadi, R. J. Dunn, S. D. Gribble, H. M. Levy, "An Analysis of Internet Content Delivery Systems", *ACM SIGOPS Operating Systems Review*, Winter 2002, Vol. 36, pp. 315 - 327.

[85] W. Kelly, P. Roe, "A Framework for Automatic and Secure Cycle Stealing", in *Proceedings of the 7th International Conference on High Performance Computing and Grid in Asia Pacific Region*, July 2004, pp. 74 - 80.

[86] K. Sripanidkulchai, "The Popularity of Gnutella Queries and its Implications on Scalability", in *Proceedings of the O'Reilly P2P and Web Services Conference,* September 2001.

[87] T. Hamada, K. Chujo, T. Chujo, X. Yang, "Peer-to-Peer Traffic in Metro Networks: Analysis, Modeling, and Policies", in *Proceedings of IFIP/IEEE Network Operations and Management Symposium (NOMS 2004)*, Seoul, Korea, April 2004, Vol.1, pp. 425 - 438.

[88] P. Kirk, "Gnutella 0.6 - Defining a Standard", July 2003, available on-line: `http://rfc-gnutella.sourceforge.net/index.html`.

[89] M. Srivatsa, B. Gedik, L . Ling, "Scaling Unstructured Peer-to-Peer Networks with Multi-tier Capacity-aware Overlay Topologies", In *Proceedings of the* 10th *International Conference on Parallel and Distributed Systems (ICPADS 2004)*, Newport Beach, CA, July 2004, pp. 17 - 24.

[90] A. Tagami, T. Hasegawa, "Analysis and Application of Passive Peer Influence on Peer-to-Peer Inter-domain Traffic", in *Proceedings of the 4th International Conference on Peer-to-Peer Computing,* Zurich, CH, August 2004, pp. 142 - 150.

[91] C. Kattirtzıs, E. Varvarigos, K. Vlachos, G. Stathakopoulos, M. Paraskevas, "Analyzing Traffic Across the Greek School Network", in *Proceedings of the 14th IEEE Workshop on Local and Metropolitan Area Networks (LANMAN 2005)*, Chania, Crete, September 2005, 18 - 21.

[92] R. Schollmeier, G. Schollmeier, "Why Peer-to-Peer (P2P) Does Scale: an Analysis of P2P Traffic Patterns", in *Proceedings of the 2nd International Conference on Peer-to-Peer Computing,* (P2P 2002), Linköping, Sweden, September 2002, pp. 112 - 119.

[93] Z. Ge, D. R. Figueiredo, S. Jaiswal, J. Kurose, D. Towsley, "Modeling Peer-to-Peer File Sharing Systems", in *Proceedings of IEEE INFOCOM03*, San Francisco, CA, April 2003, Vol. 3, pp. 2188 - 2198.

[94] C. Shannon, D. Moore, K. C. Claffy, "Beyond Folklore: Observations on Fragmented Traffic", *IEEE/ACM Transactions on Networking*, Vol. 10, No. 6, pp. 709 - 719, December 2002.

[95] J. Poulwelse, J. Garbacki, P. Epema, D. Andsips, "The BitTorrent P2P File-sharing System: Measurements and Analysis", in *Proceedings of the 4th International Workshop on Peer-to-Peer Systems (IPTPS 2005),* Ithaca NY, February 2005.

[96] D. Qiu, R. Andsrikant, "Modeling and Performance Analysis of BitTorrent-like Peer-to-Peer Networks", in *Proceedings of the ACM SIGCOMM,* Portland, OR, August 2004, pp. 367 - 378.

[97] Former eDonkey 2000 website (now reporting a copyright warning): `http://www.edonkey2000.com/`.

[98] K. Tutschku, "A Measurement-based Traffic Profile of the eDonkey Filesharing Service", in *Proceedings of the 5th International Workshop on Passive and Active Network Measurement (PAM 2004)*, Antibes Juan-les-Pins, France, April 2004, pp. 12 - 21.

[99] T. Hossfeld, K. Tutschku, F.U. Andersen, "Mapping of file-sharing onto Mobile Environments: Feasibility and Performance of eDonkey with GPRS", *Wireless Communications and Networking Conference, 2005 IEEE*, Vol.4, pp. 2453 - 2458, March 2005.

[100] T. Hossfeld, K. Tutschku, F.U. Andersen, "Mapping of file-sharing onto Mobile Environments: Enhancement by UMTS", in *Proceedings of the 3rd IEEE International Conference on Pervasive Computing and Communications Workshops, PerCom 2005*, pp. 43 - 49, March 2005.

[101] D. Stutzbach, R. Rejaie, S. Sen,"Characterizing Unstructured Overlay Topologies in Modern P2P File-Sharing Systems", *Internet Measurement Conference 2005*, Berkeley, CA, October 2005.

[102] R. Pang, V. Paxson, R. Sommer, L. Peterson, "BinPac: A YACC for Writing Application Protocol Parsers", in *Proceedings of ACM SIGCOMM, Internet Measurement Conference*, Pisa, Italy, October 2006.

[103] R. Bolla, M. Canini, R. Rapuzzi, M. Sciuto, "On the Double-Faced Nature of P2P Traffic", in *Proceedings of the 16th Euromicro Conference on Parallel Distributed and Network-Based Processing (PDP 2008),* pp. 524 - 530, Toulouse, France, February 2008.

[104] The mySQL Open Source Database, official homepage: `http://www.mysql.org`

[105] M. Lutz, D. Ascher, "Learning Pyhton", Second Edition, December 2003, O'Reilly, ISBN: 0-596-00281-5.

[106] CAIDA "NetGeo - The Internet Geographic Database", available on-line at: `http://www.caida.org/tools/utilities/netgeo/`.

[107] C. Davis, P. Vixie, T. Goodwin, I. Dickinson, "A Means for Expressing Location Information in the Domain Name System", *IETF, RFC* **1876**, January 1996, available on-line at: `http://www.ietf.org/rfc/rfc1876.txt?number=1876`.

[108] M. Feldman, C. Papadimitriou, J. Chuang, I. Stoica, "Free-Riding and Whitewashing in Peer-to-Peer Systems", in *Proceedings of the ACM SIGCOMM Workshop on Practice and Theory of Incentives in Networked Systems (PINS),* Portland OR, August 2004, pp. 228 - 236.

[109] K. Lai, M. Feldman, I. Stoica, J. Chuang, "Incentives for Cooperation in Peer-to-Peer Networks", 1st Workshop on Economics of Peer-to-Peer Systems, June 2003, Berkeley CA.

[110] W. Aiello, F. Chung, L. Lu, "A Random Graph Model for Massive Graphs", in *Proceedings of the* 32nd *Annual ACM Symposium on Theory of Computing,* Portland, OR, ACM Press, 2000, pp. 171 - 180.

[111] L. Bresalu, P. Cao, L. Fan, G. Phillips, S. Shenker, "Web Caching and Zipf-like Distributions: Evidence and Implications", in *Proceedings of IEEE INFOCOM* 1999, New York, NY, March 1999, Vol. 1, pp. 126 - 134.

[112] K. Gummadi, P. Dunn, R. J. Saroiu, S. Gribble, S. D. Levy, H. M., J. Zahorjan, "Measurement, Modeling, and Analysis of a Peer-to-Peer File-sharing Workload", in *Proceedings of the 19th ACM Symposium on Operating Systems Principles (SOSP 2003),* Bolton Landing, NY, October 2003; ACM Press, New York, NY, pp. 314 - 329.

Index

A

B

C

D

E

F

G

H

I

J

K

L

M

N

O

P

Q

R

S

T

U

V

W